Personalizing
Precision Medicine

Personalizing Precision Medicine

A Global Voyage from Vision to Reality

Kristin Ciriello Pothier

Boston, MA, USA

Registered Office
John Wiley & Sons, Inc., 111 River Street, Hoboken, NJ 07030, USA

Editorial Office
111 River Street, Hoboken, NJ 07030, USA

For details of our global editorial offices, customer services, and more information about Wiley products visit us at www.wiley.com.

Wiley also publishes its books in a variety of electronic formats and by print-on-demand. Some content that appears in standard print versions of this book may not be available in other formats.

Library of Congress Cataloging-in-Publication Data

Names: Pothier, Kristin Ciriello, 1974– author.
Title: Personalizing Precision Medicine: A Global Voyage from Vision to Reality / Kristin Ciriello Pothier.
Description: Hoboken, NJ : John Wiley & Sons, Inc., 2017. | Includes bibliographical references and index. |
Identifiers: LCCN 2017020345 (print) | LCCN 2017021066 (ebook) | ISBN 9781118792179 (pdf) | ISBN 9781118792124 (epub) | ISBN 9781118792117 (pbk.)
Subjects: | MESH: Precision Medicine–trends
Classification: LCC R733 (ebook) | LCC R733 (print) | NLM WB 300 | DDC 610–dc23
LC record available at https://lccn.loc.gov/2017020345

Cover Design and Image: Courtesy of Diana Saville.

Set in 10/12pt Warnock by SPi Global, Pondicherry, India

Printed in the United States of America

10 9 8 7 6 5 4 3 2 1

Contents

Acknowledgments

This book would not have been completed without the support of my loving family—Bryan, Olivia, and Luke—and our parents and dear friends and the support of Ernst & Young LLP. I would like to give special thanks to my *Personalizing Precision Medicine* life sciences lead editorial team in the firm's Parthenon-EY practice, led by Mahala Burn, Brian Quinn, Joe Zaccaria, Jessica Lin, and Jay Canarick. Additional thanks goes to our team of Parthenon-EY consultants who contributed to the research in the book, including Alasdair Milton, Jay Buckingham, Jessica Bernheim, Alex Chen, Glenn Engler, Will Janover, Ankit Goel, Eric Haskel, Ryan Juntado, Hayley Kriman, Harish Kumar, Chen Liu, Melissa Maggart, Armelle Sérose, Hanu Tyagi, Eleonora Brero, Shushant Malhotra, Derek Matus, Amy McLaughlin, Will Poss, Melanie Gaynes, Chris Bravo, Gillian O'Connell, Jeremy Rubel, Hamza Sheikh, Mark Sorrentino, Yuan Wang, Maxine Winston, Gary Yin, Ike Zhang, and Scott Palmer. Also, thanks for the support of my Ernst & Young leadership teams worldwide, my quality, legal, graphics, and marketing teams; my cover illustrator Diana Saville; and external copy editor and oldest, most honest friend, Kristin Walker Overman. Also incredible thanks to my publisher Wiley and its team led by Jonathan Rose for having the patience to wait for the completion of this book. And finally, thanks to my friends, family, and trusted colleagues affected by cancer, who generously devoted their time and support to help me make this book a reality.

Introduction

It was the spotted lung scan day.

My old friend Heather[1] said this to me, matter of fact, over pizzas one evening as she described one of many pivotal moments in the last 2 years of her life with cancer.

Her statement, on a day memorialized in her mind by a diagnostic test, is a small window into the life of a cancer patient. Her life pre-cancer was busy, successful. She managed her own design business and ran hard and spirited, traits we were all envious of in college and that stuck with her. There was no time for anything other than her business, her husband, and her zest for life. A nagging, mild tender breast that just didn't go away prompted her to go to the doctor more out of annoyance than anything else. She was slated to meet her husband abroad at the end of the week for a European vacation; this was 1 appointment out of 20 she needed to check off her list. She hadn't even had a routine mammogram because she had barely reached her 40s and hadn't gotten around to it yet. That European trip never came.

Heather watched curiously as the administrator of her mammogram went from calm to concerned to alarmed. The administrator called her boss. Her boss's face also did not hide the alarm. Heather was scheduled for a biopsy immediately.

The days after the diagnosis were filled with research and calls to any friends that could help, from a medical point of view and from a support point of view. Thankfully, Heather's support system also included a childhood friend who was a cancer researcher at one of the top institutions in the United States and could get her to the right physicians in her own city.

And then the treatment began.

Her days, which before were measured in her meetings with clients to restore their historic homes, her long dinners with her husband, and her energetic throes into SoulCycle, were now measured differently. Days measured in test

---

1 All previously unpublished patient names have been changed.

results, in exhaustion from the drug regimens, in pain from the surgeries, and in desperate hope for the next day highlighted in her support systems surrounding her. She did everything right. Her physicians personalized her treatment for her with drugs that would work on her specific cancer. The cancer looked like it was gone. And then, a follow-up scan suggested a suspected relapse and potential spread of the cancer to the lung. Her lungs looked, well, "spotted." And it started all over again.

Indeed, as much as this experience was personal to Heather, we are only starting to personalize cancer treatment in the truest sense. "I did certain things to 'hack' my experience specifically for me ... both medically and psychologically. But I understand it is difficult for doctors to do this with thousands of patients," Heather said. The ability to tailor a drug regimen to a specific genetic code that is truly personalized to that specific DNA double helix has been a dream of researchers, physicians, and patients alike. Advances in precision medicine, specifically around the genome and the helices embedded, are making this dream a reality.

Patients struggle with "chemo," drugs that indiscriminately kill both cells in their tumors and normal cells like their hair follicles, the lining of their throats and stomachs, and their sperm and eggs in their reproductive systems. According to Christopher Cutie, a urologist by training and the current chief medical officer of the innovative bladder cancer company Taris, "With any therapy that hasn't been tested for a lifetime of a patient, there is risk of what the body may do. The body craves homeostasis. When we expose it to insult, even if correcting one part of the body, it may manifest itself differently in another part."

Today, biomarkers directly connected to drugs or to crucial outcomes in the human body allow physicians to identify drugs that are most likely to help a patient, and those drugs can be used to target cancerous cells only, which reduces the side effects that the patient experiences. 28% of all drugs approved by the FDA today have biomarker information, with more in the pipeline to come. But while new advances in precision medicine hold so much promise, many challenges must be overcome before precision medicine can truly transform healthcare. For example, former President Obama's Precision Medicine Initiative aims to collect genetic and metabolomic profiles, medical records, and other health information for at least one million people, and the wealth of data will help researchers advance their understanding of diseases. "Wearables," which are devices like watches or chest monitors worn on a person, will also aid in the collection of tremendous amounts of health data. However, fundamental questions must first be addressed, such as how to store these sensitive data, how to share the data, and how to use these data to create value for patients.

Furthermore, access to healthcare remains a global challenge. Targeted therapies are among the most sought-after and most expensive therapies in the world, and market access and payment issues must be solved to ensure that precision medicine benefits all patients, not just a select few. Here in the United States, we have built some of the most prestigious cancer centers in the world, and the likes of the University of Texas MD Anderson Cancer Center, Memorial Sloan Kettering Cancer Center, Massachusetts General Hospital, and Mayo Clinic provide some of our best demonstrated examples of precision medicine from vision to reality. But not all regions of the world, or even regions of the United States, have complete access to these types of institutions although they each continue to enhance and broaden their reach, and I admittedly spend more time in this book analyzing the global challenges we are facing when access is not yet achieved.

The power of precision medicine also opens the door to controversy given that the most advanced techniques can be used to do far more than cure disease. Many fear that new technology will enable the creation of designer babies or the elimination of diversity. While many scientists have discussed limitations on this type of human engineering, biomedical research is global, and there is no single authority that can limit how technologies are used.

This book explores the advances that have been made in precision medicine and discusses the global implications for companies, payers, researchers, physicians, and patients who are translating precision medicine from vision to reality. The research and one-on-one discussions with pioneers in precision medicine, day-to-day caregivers, and patients and their supporters worldwide provide firsthand experience into the reality behind the hype and demonstrate the raw emotion in building an entirely new discipline that not only brings so much good to our patients in need but also introduces many challenges. We have truly hit a new frontier, and the goal here is to bring clarity to the progress we have made, to begin a discussion of the complexities and challenges we face, and to inspire hope in the future we are building by **personalizing precision medicine**.

Methodology

This book is based on my experience in working in precision medicine strategy for products and services across diagnostics, life sciences, and therapeutics companies, investor groups, and medical institutions for over 20 years, extensive secondary research, and over 100 primary interviews with key stakeholders worldwide.

The secondary research included drug pipeline research to uncover both current and future precision medicine drugs and the diagnostics that fuel them, scientific and clinical literature reviews on existing and emerging technologies within precision medicine, and website searching to verify the most cutting edge products, services, and offerings that fuel this industry.

The primary research included detailed one-on-one interviews with industry executives, laboratorians, physicians, payers, patients, and their caregivers in the United States, Europe, South America, Central America, the Caribbean, India, China, Japan, and the Middle East who gave their feedback, insights, and detailed views in order to promote education of precision medicine and to show the diverse impact of precision medicine among a range of stakeholders and regions around the globe.

About the Author

Kristin Ciriello Pothier is the global head of Life Sciences for the Parthenon-EY practice of Ernst & Young LLP. She has over 20 years of experience in business strategy and medical research in the life sciences industry. She is a noted international speaker, workshop leader, and writer in life sciences. She is also a clinical laboratory and medical innovation expert, helping develop and implement product and service strategies worldwide for investors, corporations, and medical institutions. Prior to EY, Kristin was a partner at Health Advances, a healthcare consulting company, and a research scientist and diagnostics developer at Genome Therapeutics, a commercial company sequencing for the Human Genome Project and at Genzyme, developing pioneering noninvasive prenatal tests and numerous other precision medicine-based diagnostics tests and algorithms. She earned an undergraduate degree in biochemistry from Smith College and a graduate degree in epidemiology, health management, and maternal and child health from the Harvard School of Public Health. She is also a founding director of BalletNext, a ballet company based in New York celebrating the convergence of innovative dance, music, and art. Kristin lives in Massachusetts with her husband and their two lively children.

Part 1

The History

1

The Right Drug, the Right Patient, the Right Time

Foundations of Precision Medicine

Eight-year-old Caleb Nolan faced an uncertain future when he was born. At 3 weeks old he was diagnosed with cystic fibrosis (CF), an inherited, devastating, incurable genetic disease that causes a buildup of mucus in various organs, including the lungs, pancreas, liver, and intestines. This results in poor weight gain, infertility, and chronic lung infections that can lead to respiratory failure. While antibiotics are used to treat infections, many CF sufferers eventually require a lung transplant, and few used to live beyond the age of 50. However, Caleb now lives a full life and will probably die of old age instead of CF [1].

CF is caused by an abnormality in the CF transmembrane regulator (CFTR) gene that prevents the normal movement of chloride ions across membranes and currently afflicts about 30,000 children and adults in the United States and 70,000 people worldwide [2]. As with all people, CF patients have two copies of the CTFR gene, one from each parent, but for them both copies are harmfully mutated. There are people who are "carriers" for the mutation who have one normal and one mutant copy, and while they do not have symptoms of CF, they can pass the gene to their children. At this moment, there are about 10 million carriers in the United States.

One of the most common mutations causing CF is called the deltaF508 deletion mutation, which can be detected using a number of molecular techniques. Akin to an error in the blueprints for building a cabinet that misses a shelf support, this means that the CTFR protein made from the blueprint of the mutated CTFR gene is defective due to a deletion at the 508th place along the protein code (see Figure 1.1). This kind of mutation can be tested by DNA amplification—making many copies of parts of someone's DNA CFTR gene and looking for the mutations that cause CF. There are three ways to assess CF: "carrier screening" of parents-to-be or pregnant women determines their CFTR gene status and helps families adequately prepare for the results. Testing is also done on amniocentesis samples to directly assess CF status in the unborn child. Finally, newborns are screened in all 50 states to assess CF status [4].

Personalizing Precision Medicine: A Global Voyage from Vision to Reality, First Edition.
Kristin Ciriello Pothier.
© 2017 John Wiley & Sons, Inc. Published 2017 by John Wiley & Sons, Inc.

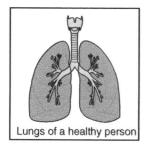

Lungs of a healthy person

Normal CFTR sequence				
Nucleotide: ATC	ATC	TTT	GGT	GTT
Amino acid: Ile	Ile	**Phe**	Gly	Val
Position: 506		**508**		510

Deleted in Delta F508

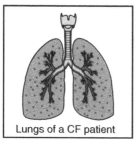

Lungs of a CF patient

Delta F508 CFTR sequence			
Nucleotide: ATC	ATC	GGT	GTT
Amino acid: Ile	Ile	Gly	Val
Position: 506			

Figure 1.1 The CFTR defect in cystic fibrosis. Source: Data from Pothier et al. [3].

Did You Know?

DNA stands for *de*oxyribo*n*ucleic *a*cid, and it is the hereditary material in humans and virtually all other organisms. The information that DNA carries is stored as a "code" that is made of four different chemicals called bases—adenine (A), guanine (G), cytosine (C), and thymine (T). The sequence of the bases dictates how that organism is made and, although there are around three billion bases in humans, the sequence is 99.9% identical.

DNA bases pair up, with adenine (A) pairing with thymine (T) and guanine (G) pairing with cytosine (C). A sugar molecule and a phosphate molecule also attach to each base to create a nucleotide, and these nucleotides are then arranged into the DNA double helix (see Figure 1.5). The DNA is arranged into 46 structures called chromosomes that are arranged into 23 matching pairs, which include one sex pair. A female has 2 X's as the sex pair, and a male has an X and a Y. The entire collection of 46 chromosomes is called a karyotype.

DNA is made of building blocks called nucleotides, and as a highly dynamic and adaptable molecule, it can suffer many types of mutations. Some are harmless, some are helpful, and some can harm the DNA and, ultimately, the organism. These mutations can arise by chance or through environmental factors such as exposure to chemicals and can occur in somatic cells (nonreproductive cells) or germ cells (reproductive cells such as sperm and egg). Diseases like cancer largely affect somatic cells, while mutations in germ cells lead to inherited diseases like cystic fibrosis. Various types of mutation exist.

There are nucleotide substitutions, where one nucleotide is swapped for another; insertions and deletions (indels), where nucleotides are added or deleted; and frameshift mutations, which are insertions or deletions of more than one nucleotide that result in the complete alteration of the sequence of a protein [6].

However, CF sufferers have new hope for treatment. This is due to a revolutionary treatment approach that was approved by the FDA in 2012. The drug Kalydeco is the first to target the underlying genetic cause of the disease. Kalydeco is one of a new generation of medicines that are specifically tailored to treat a disease based on the genetic makeup of an individual. In the case of Kalydeco, rather than just treating the symptoms of the disease, the drug acts on the gating defect associated with the defective CFTR protein, helping to open up the blocked chloride channels (see Figure 1.2). This allows for a clearing of the mucus buildup from the inside out. These drugs are part of a new age in medicine, "precision medicine," which strives to provide the right treatment for the right patient at the right time.

Sick CF patient

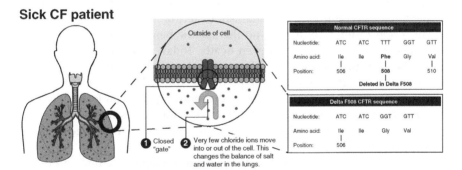

CF patient treated with Kalydeco

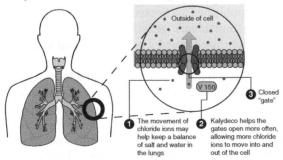

Figure 1.2 Kalydeco's mechanism of action. The drug acts to open up the protein gate that regulates the movement of chloride ions between the outside and the inside of the cell. Source: Adapted from Kalydeco [7].

The Origins of Precision Medicine

The 1990s was a golden age for the pharmaceutical industry. This was the original "blockbuster" era, when the focus was on developing broad-spectrum drugs for large "primary care" indications like high cholesterol, asthma, and depression. Adopting a "one-size-fits-all" approach meant that companies could generate billions of dollars in sales by targeting the largest patient populations, spending hundreds of millions of dollars on marketing campaigns, and having large sales forces focusing on doctors. Even drugs that were fourth, fifth, or sixth to market could achieve stellar sales performance using this approach, with insurance companies willing to cover these products.

Drug development strategies used by many companies during the 1990s were crude and rudimentary compared with those commonly utilized today. When hunting for a new multibillion-dollar drug, companies screened huge libraries of compounds against a target to look for a suitable candidate that was then tested in clinical trials. The companies searched without assessing whether certain people would be nonresponders or would have an adverse reaction. In some cases, researchers didn't even fully understand the compound's mechanism of action.

This often resulted in little or no therapeutic benefit and, in some cases, caused serious side effects and even death. Today, less than half of people prescribed an antidepressant achieve remission with the first therapy [8], while patients treated for asthma, type II diabetes, arthritis, and Alzheimer's all have differing responses to their medications [1] that can lead to limited therapeutic outcomes or serious side effects. Overall, it is estimated that many of the leading drugs in the United States today only benefit between 1 in 25 and 1 in 4 patients [9].

Warfarin, a common blood thinner, can cause major bleeding and death due to the fact that patients respond to the drug in different ways, driven by genetic variations on a person-to-person basis. However, an analysis of a number of independent studies published in September 2014 showed that dosing of warfarin based on an individual's genetics could reduce major bleeding episodes by over 50%, pointing toward a personalized approach to treatment with the drug [10].

The rigid drug development approach of the past has resulted in the termination of a number of compounds late in the clinical trial process, as companies fail to find a therapeutic signal or, worse still, uncover a major safety issue. On many occasions, this is due to the heterogeneous nature of the patient populations that are used in the trials, with researchers never fully understanding the genetic, environmental, or lifestyle factors that can influence drug response in these large population-wide studies [9].

The impact of this approach to drug development was highlighted in 2005 when Tysabri, an immunosuppressant drug used to successfully treat multiple sclerosis, was removed from the market following three cases of a rare neurological condition called progressive multifocal leukoencephalopathy (PML). Two patients subsequently died [11].

Tysabri was returned to the market in 2006 with a black box warning—a statement of serious risks required by the FDA—and a risk management plan. As part of this, the drug's manufacturer, Biogen Idec, worked with a lab to develop a test that helped stratify patient risk based on the specific presence in the patient's body of the John Cunningham (JC) virus, which can cause PML in patients with compromised immune systems. Biogen had therefore developed a precision approach to treatment with Tysabri.

These examples highlight the complex, multifactorial nature of drug response. Often, biomarkers or molecular pathways that scientists think are involved in disease turn out only to be associated with the disease, rather than to be the root cause (see Figure 1.3). The result can be billions of dollars in wasted R&D spend and drugs that either have limited therapeutic benefit or, in worse case scenarios, can actually cause serious harm to patients. This is why precision medicine promises to be revolutionary for the field of medical science.

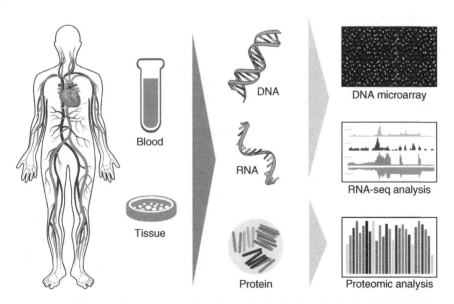

Figure 1.3 Molecular diagnostics examine the molecules in the cell, that is, the DNA, RNA, or proteins, and their role in human biology and disease. Source: Data from Pothier et al. [3].

The idea of tailoring medicine to an individual is not new. Hippocrates, the "Father of Western Medicine," said that "it is more important to know what sort of person has a disease, than to know what sort of disease a person has." [12] However, it would be another 2500 years before precision medicine became a reality.

The scientific underpinnings of precision medicine began in the late 1860s when Swiss chemist Friedrich Miescher stumbled across a new molecule as he was trying to isolate proteins from white blood cells. Instead of successfully isolating proteins, he discovered a substance in the cell nucleus that had chemical properties that were different from those of proteins.

He named this new molecule "nuclein," deducing that it consisted of hydrogen, oxygen, nitrogen, and phosphorus, and, believing that he had discovered something important, he stated that "it seems probable to me that a whole family of such slightly varying phosphorous-containing substances will appear, as a group of nucleins, equivalent to proteins" [13]. Miescher, who was largely forgotten after his death, had discovered DNA (Figure 1.4).

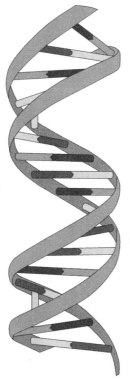

1902
Emil Fischer shows that amino acids are linked and form proteins

1929
Phoebus Levene discovers deoxyribose sugar in nucleic acids

1944
Oswald Avery, Colin McLeod, and Maclyn McCarty show that DNA, and not protein, was the hereditary material of bacteria

1952
Maurice Wilkins and Rosalind Franklin create an X-ray image of DNA

1961
Marshall Nirenberg shows that three nucleotides code for an amino acid, thereby cracking the genetic code

1983
Kary Mullis invents PCR

1987
The first automated DNA sequencing machine is introduced

1990
The Human Genome Project is announced

1869
Friedrich Meischer discovers DNA and names it nuclein

1911
Thomas Hunt Morgan shows that genes are linearly located along chromosomes

1941
George Beadle and Edward Tatum discover that genes make proteins

1950
Edwin Chargaff finds that cytosine complements guanine and adenine complements thymine

1953
James Watson and Francis Crick reveal DNA to be a 3D helical structure

1970s
DNA sequencing is invented

2003
The human genome is completed

Figure 1.4 The history of DNA.

Russian scientist Phoebus Levene, Austrian biochemist Erwin Chargaff, American scientist Oswald Avery, and English chemist Rosalind Franklin played major parts in linking Miescher's original discovery of nuclein in 1869 to the 1953 announcement by James Watson and Francis Crick that DNA exists as a three-dimensional double helix (Figure 1.5). Watson and Crick won the Nobel Prize in Physiology or Medicine in 1962 for their discovery. Sometimes forgotten is their colleague in this research, Rosalind Franklin, a chemist who also made the discovery with the team but died of ovarian cancer before the Nobel Prize was awarded. Unfortunately, Nobel Prizes are not given posthumously.

Although they had now discovered the "blueprint" for a human being, scientists in the 1950s still didn't know how the information held in our DNA was translated to the 20-letter alphabet of amino acids, the building blocks of proteins, which are the functional units that ultimately drive cellular processes. In 1939, the role of another nucleic acid, RNA, had been linked to protein synthesis, but it wasn't until the 1950s that the various types of RNA that are

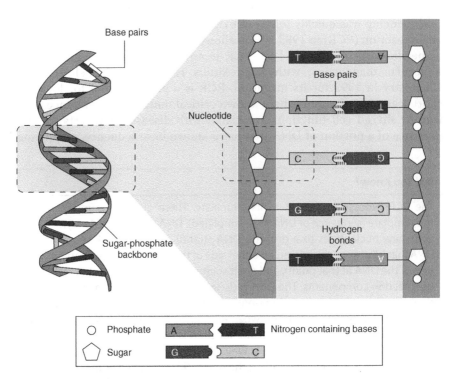

Figure 1.5 The double-helix structure of DNA. The molecule consists of four bases—adenine (A), guanine (G), cytosine (C), and thymine (T) and a sugar-phosphate backbone. Adenine always binds to thymine (A–T) and guanine always binds to cytosine (C–G). Source: Adapted from Encyclopædia Britannica [5].

essential in turning the DNA code to protein were identified. It's now known that when DNA is replicated, it's translated to single-stranded messenger RNA (mRNA), with the thymine base (T) replaced with uracil (U). The process of converting the mRNA to protein happens on a molecule called a ribosome. Another type of RNA called transfer RNA (tRNA) can bind to the free amino acids and bring them to the ribosome, where the tRNA reads the mRNA code and starts to build out the protein. In his groundbreaking paper in 1961, Marshall Nirenberg presented results from an experiment that showed that three nucleotides coded for an amino acid. With this discovery, the genetic code had finally been cracked (see Figure 1.6) [15].

In the 1970s, a technique was invented that would revolutionize the field of molecular biology and would prove to be essential to the future of precision medicine—the development of DNA sequencing. Until the advent of DNA sequencing, molecular biologists could only examine DNA indirectly through protein or RNA sequencing [16]. What scientists lacked was the ability to sequence, analyze, and interrogate the building blocks of life in order to locate gene sequences and identify mutations in the genetic code.

Sequencing was quickly followed by the introduction of the polymerase chain reaction (PCR) in 1983, which allowed researchers to quickly amplify a specific target sequence [17]. PCR is now considered a workhorse in molecular diagnostics, with Kary Mullis receiving a Nobel Prize in Chemistry in 1993 for its invention. PCR is a powerful tool for locating short segments of a gene where known critical mutations or variances can lead to altered cell functions associated with disease. PCR tests for the presence of a portion of DNA that has a known base sequence, employing

---

Did You Know?

The PCR process is elegant in its simplicity. There are four components— the template (sequence of DNA to be amplified), DNA polymerase (enzyme that adds new nucleotides to a growing DNA strand), primers (small segments of DNA that bind a specific region on either side of the target DNA and start replication of the DNA at that point), and a salt solution called the buffer that stabilizes the reaction components. The DNA is denatured (hydrogen bonds holding the double helix are broken, creating single-stranded DNA) by heating to more than 90°C. As the mixture cools to between 40 and 60°C, the primers bind to their target sequence on the template. The reaction is then heated to around 72°C, which is the optimal temperature at which DNA polymerase operates. The polymerase extends the primers, adding the nucleotides onto the primer in the correct order, based on the sequence of the template. This process is then repeated over and over again. Because the DNA made in the previous cycle can also serve as a template, the resulting amplification of DNA is exponential [3].

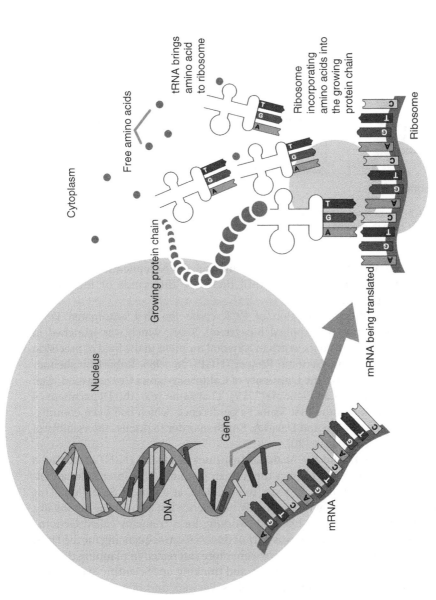

Figure 1.6 From DNA to protein. Single-stranded mRNA is created when DNA is replicated. The mRNA is then translated to protein on the ribosome, where tRNA reads the three-letter code (or codon) in the mRNA, with each codon coding for a different amino acid in the protein being built. Source: Adapted from "What is protein synthesis" [14].

Amplification cycles

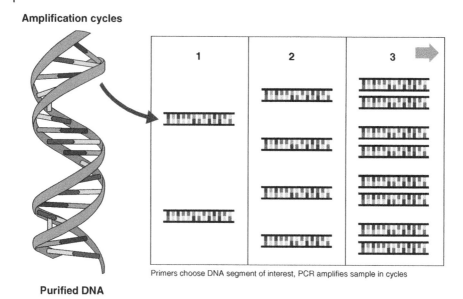

Primers choose DNA segment of interest, PCR amplifies sample in cycles

Purified DNA

Figure 1.7 The polymerase chain reaction (PCR) process. PCR amplifies DNA by repeating cycles that each duplicates the base sequence in a specific section of the DNA strand [18].

the same enzymatic process used by natural DNA replication to rapidly amplify, or copy, that sequence until there are thousands or millions of copies (see Figure 1.7) [4].

Shortly after the introduction of PCR came the first automated DNA sequencing machine, the Applied Biosystems 370A, which was launched in 1987. This system proved its worth in a pivotal moment in the field of precision medicine—the Human Genome Project (HGP). In 1985, Robert Sinsheimer organized a workshop at the University of California–Santa Cruz entitled "Can We Sequence the Human Genome?" [19]. That same year, the Department of Energy (DOE) funded the first Santa Fe conference, which had been commissioned by Charles DeLisi and David A. Smith in order to discuss the viability of a human genome initiative [20].

In 1990, the DOE and the National Institutes of Health (NIH) presented a plan to the US Congress to map the human genome. This international collaboration eventually took 13 years to complete, at a cost of $3 billion. This time in history is a special time not only for science but also for me personally, as my husband and I met while both working for a company called Genome Therapeutics, a company that was part of the US team sequencing for the HGP. Admittedly, early in our careers, we were more interested in planning the lab's parties, throwing around our pipettes, and thinking up "Scientific Word of the Day" than making history, but nevertheless, we were part of it.

Although it was called the Human Genome Project, scientists also aimed to sequence a number of other organisms, including mouse, worm, fruit fly, yeast, and *bacteria*. The race heated up in 1998 when a private company, Celera Genomics, announced its intention to sequence the human genome using a method called whole genome sequencing (WGS). WGS allows for the complete genome of an organism to be sequenced in one single step, utilizing newer, automated sequencing approaches such as next-generation sequencing (NGS), rather than the manual methods, called Maxam–Gilbert and Sanger sequencing. In 2001, draft versions of the entire human genome sequence were published from both the HGP and Celera. Although the race had been a tie, Celera's WGS approach had, in 3 years, achieved what the HGP had in 11 years, at a cost of $300 million, one-10th the cost of the publically backed initiative [20, 21].

Importantly, the Human Genome Project also spurred the introduction of a number of NGS systems from now defunct company 454, Life Technologies (now Thermo Fisher), and Illumina, which were able to produce greater amounts of data, faster, and more cheaply than the Sanger-based systems. While these NGS platforms have continued to evolve, new "third-generation" platforms (3GS) from PacBio and Oxford Nanopore can now produce DNA reads that are thousands of bases long rather than the hundreds of bases produced by NGS. This produces a trade-off. NGS produces more accurate reads compared with 3GS, but the increased read length from 3GS is less burdensome on researchers when the pieces of the DNA jigsaw need to be put back together. However, NGS remains the mainstay today due to its lower cost and higher accuracy.

Another equally important event in the development of DNA technologies occurred in the 1990s but hardly registered at all due to the excitement around the HGP and the invention of automated DNA sequencing—the introduction of the DNA microarray. First invented in the late 1980s by Stephen Fodor and his colleagues at Affymax, DNA microarrays (or DNA chips) are used to measure gene expression levels. The technique uses the complementary nature of nucleic acid binding (A to T and G to C) to "probe" for fluorescently labeled "targets." The DNA probe is attached to a solid surface such as glass or silicon, and the labeled target is added. The intensity of the signal from the bound targets can then be compared with a control, thereby allowing relative quantification of the signal under different conditions.

Microarrays (both DNA and RNA) allow researchers to quantify gene expression, determine binding sites of DNA transcription factors (proteins that bind DNA and help turn it into mRNA), and, crucially, identify single nucleotide polymorphisms (SNPs), which are single nucleotide changes in a gene, often associated with disease and drug response [22, 23]. Although NGS and 3GS technologies have supplanted microarray in many respects, it remains a key part of the molecular biologists' toolkit, in large part due to its lower cost compared with NGS.

Technology Toolkit in Place: Transition to Meaning in the Clinic

In reading this chapter, you may by now have celebrated the glory of science, checked the prize money for a Nobel Prize considering so many of these scientists seemed to have won one, or fallen fast asleep. The technological underpinnings of genetics have that effect on people. But the exciting part is yet to come. These researchers had developed the ability to look into the molecular structure of a human and/or its tumor and identify its characteristics at the core, paving the way to use these characteristics to manipulate the genome and course correct it for our patients. The transition of these technologies into the clinic was occurring at the same time.

Let's talk about one of those stories in creating a game-changing drug for women with breast cancer: Herceptin. The first step in Herceptin's story began in the early 1980s, nearly 20 years before its launch, when scientists identified a gene called *HER2* that was related to epidermal growth factor receptors (EGFRs) [24], surface proteins involved in a range of biological processes, including cell proliferation (an increase in the number of cells as a result of cell growth and division), angiogenesis (formation of blood vessels), and inhibition of apoptosis (cell death).

It was subsequently discovered that the *HER2* gene was mis-regulated in up to 25% of breast cancers, resulting in too many copies of the resulting protein and contributing to poor prognosis for these patients [24]. As a molecule that was directly implicated in the development of this particular form of breast cancer, the HER2 protein was a biological marker, or biomarker, for the disease.

Biomarkers are measurable indicators of healthy and pathological processes and can be derived from various types of molecules, including body DNA, RNA, protein, or lipids [25]. The World Health Organization defines a biomarker as "any substance, structure, or process that can be measured in the

---

Did You Know?

In a very simple analogy, cancer can be thought of as a disease of the gas pedal and the brake in a car. Cancer results when cells divide uncontrollably, moving (metastasizing) to other parts of the body, where they grow and grow until they put pressure on the surrounding tissues. Certain proteins called tumor suppressors (the brakes) prevent the uncontrolled cell division, but other proteins, "oncoproteins" (the gas pedal), drive the uncontrolled growth. When the genes that code for the tumor suppressors (brakes) or oncoproteins (gas) become mutated, then cancer can occur.

body or its products that influence or predict the incidence of outcome or disease" [26]. To be able to treat an individual's disease, you first have to identify specific molecular targets unique to that person's disease state. From there, you can then design a therapy that modulates that particular target.

Based on this finding, Herceptin (trastuzumab) became the first targeted therapy based on an individual's genetics when it was approved by the FDA in 1998. Herceptin targets metastatic breast cancer cells that overproduce the HER2 protein. Herceptin also became the first drug codeveloped with a test called HercepTest, which was simultaneously approved in order to aid in the identification of HER2-positive patients (see Figure 1.8) [28].

Herceptin has subsequently shown the world the power of precision medicine. In a study published in October 2014, the 10-year overall survival rate for breast cancer patients taking Herceptin in addition to undergoing chemotherapy was 84 vs. 75% for those patients treated by chemotherapy alone [29]. Although there is still some way to go before we can always deliver on the goal of "the right drug, for the right patient, at the right time," the advent of targeted drugs like Herceptin placed us firmly on a journey toward personalized medicine that began with Hippocrates some 2500 years ago.

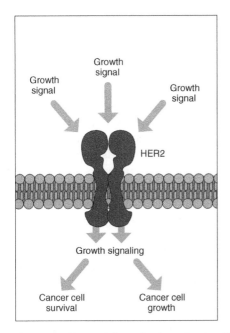
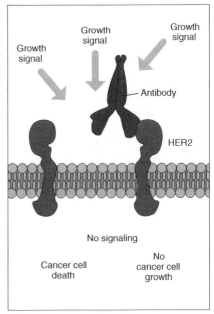

Figure 1.8 Herceptin's mechanism of action. Herceptin is a drug that binds to a cell surface protein called HER2, which is overexpressed in certain cancers. By blocking its downstream actions, Herceptin prevents HER2, causing cancer survival and growth. Source: Adapted from Nichols [27].

However, therapy selection is only one part of the precision medicine story, which spans everything from assessing an individual's risk of developing disease through screening, diagnosis, prognosis, risk assessment, therapy selection, and, finally, monitoring of possible disease recurrence. This is what we refer to as the "diagnostics continuum" (see Figure 1.9).

At the "risk" level, we can assess the likelihood of a given individual to develop a disease. For example, there are inherited mutations in the *BRCA1* and *BRCA2* genes that account for around 20–25% of hereditary breast cancers [30]. Identifying individuals with these mutations allows physicians to understand their patients' disease risk and gives more agency to patients, who can decide if they want to undertake enhanced screening or opt for prophylactic surgery.

Biomarkers can also be used to screen for disease in high-risk patients who are asymptomatic. An example is the *Early*CDT-Lung test from GeneNews, a blood-based test that measures seven autoantibodies that are known to be associated with lung cancer. The test allows physicians to screen high-risk patients such as ex-smokers in order to detect possible disease before it is visible on CT.

When disease has been detected, there are a range of biomarkers that can then provide physicians with further diagnostic insights as well as information on a patient's likely prognosis. These include diagnostic tests such as Rosetta Genomics miRview, which can differentiate between the four main subtypes of lung cancer using preoperative biopsies and prognostic tests such as the Onco*type* DX range of cancer tests from Genomic Health, which can assess risk recurrence by analyzing expression levels of a set of genes known to be associated with the disease. Prognostic markers are often helpful in avoiding unnecessary surgeries or further interventions.

As we have discussed, analysis of certain genomic biomarkers can direct therapy selection, such as in the case of *HER2* overexpression and the use of Herceptin. At the final "monitoring" stage of the continuum, the best example is in diabetes patients, who assess their blood glucose levels on an ongoing basis in order to control their disease.

As we can see, biomarkers are essential in all parts of the diagnostics continuum, but their identification and validation can be challenging given the highly complex nature of the human biological system, with more than 20,000 genes encoding for around 30,000 proteins [31].

Biomarker discovery has usually been either "hypothesis-based" or "discovery-based." The hypothesis-based model is driven by our increasing understanding of the underlying mechanisms of disease, such as in the case of diabetes, where research led to the discovery that the disease causes sustained elevation in blood glucose levels [32]. The discovery-based model is driven by the identification of changes in the presence or relative abundance of molecules that are tightly associated on a statistical basis with the disease of interest. An example of the discovery model is *BRCA1*, which was initially identified when it was mapped to a region of chromosome 17 that is frequently deleted in breast cancer [32].

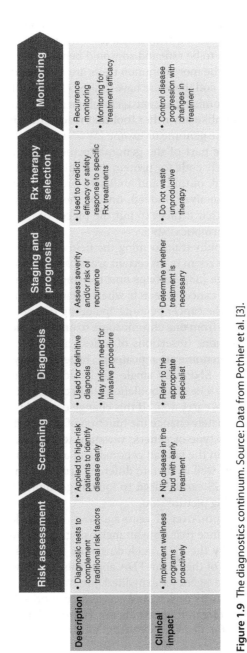

	Risk assessment	Screening	Diagnosis	Staging and prognosis	Rx therapy selection	Monitoring
Description	• Diagnostic tests to complement traditional risk factors	• Applied to high-risk patients to identify disease early	• Used for definitive diagnosis • May inform need for invasive procedure	• Assess severity and/or risk of recurrence	• Used to predict efficacy or safety response to specific Rx treatments	• Recurrence monitoring • Monitoring for treatment efficacy
Clinical impact	• Implement wellness programs proactively	• Nip disease in the bud with early treatment	• Refer to the appropriate specialist	• Determine whether treatment is necessary	• Do not waste unproductive therapy	• Control disease progression with changes in treatment

Figure 1.9 The diagnostics continuum. Source: Data from Pothier et al. [3].

Gene profiling approaches like microarrays and real-time PCR (which quantifies amplification of the target during, rather than at the end of, a PCR process) have proven useful in identifying some biomarkers but have limited predictive power. Genome-wide association studies have also been used and have yielded numerous variants associated with disease. However, many of the identified mutations have proven to be rare and so would benefit only a small number of people, while some of the more common traits are associated with only a small increase in disease risk and therefore have limited clinical utility [33].

One of the most promising recent technologies for biomarker identification is NGS, which, unlike any other tool that researchers have today, allows analysis of the complete genome or exome (WGS and whole exome sequencing). The exome is the part of the genome formed from exons (*EX*pressed regi*ONS*) that ultimately code for protein. NGS platforms continue to evolve in terms of their speed, output, and ability to produce massive amounts of data for scientists to analyze. Indeed, one of the challenges of NGS is the time, effort, and cost that it takes to put the pieces of the DNA jigsaw back together—the "bioinformatics" step.

However, NGS is rapidly transforming the field of biomarker identification as it is flexible, and can zoom in on certain areas of interest within the genome or provide full coverage [33]. Perhaps unsurprisingly its impact has been felt the most in the field of cancer research, where scientists can now assess major structural changes in the cancer genome, including translocations (the movement of pieces of DNA from one chromosome to another), deletions (deletions of DNA bases), insertions (insertions of bases), and copy number variation (entire sections of the genome are repeated). The technology can be utilized to look at SNPs as well. SNPs are single nucleotide changes within a DNA sequence. For instance, 99% of a population could carry a "C" at a certain position, while the other 1% carry a "T." If this change occurs in a coding region of the gene, it could ultimately change the function of the resultant protein. SNPs are the most common type of genetic variation found, occurring in roughly every 300 nucleotides in the human genome, and are known to be involved in a range of diseases, as well as dictate different drug responses between individuals. Therefore, NGS can be used to look at genomic variations in detail while looking at the "big picture" too.

The launch of Herceptin over 17 years ago was a landmark moment for precision medicine, and since that time an increasing number of targeted agents have been approved by the FDA, including nine new personalized drugs in 2014 alone [34]. Many of the recent FDA approvals have also taken advantage of one or more of the agency's expedited programs, which include breakthrough, fast track, priority review, and accelerated approval [35].

Among the recent targeted approvals are Harvoni for hepatitis C (2014), Cerdelga (eliglustat) for Gaucher's (2014), Vimizim for Murquio syndrome

(2014), and Repatha and Praluent both for homozygous familial hypercholesterolemia (2015).

Despite the launch of drugs in areas such as infectious, cardiovascular, and rare diseases, oncology continues to lead the way in the field of precision medicine, reflecting the strong underlying genetic drivers of the disease. This is highlighted in Figure 1.10, which shows how the number of identified mutations in lung adenocarcinoma has exploded in the last 35 years. Each of these mutations is a possible drug target.

According to the FDA, one-third of the drugs with pharmacogenomic information on the labels are for cancer [37], and a recent report from Tufts stated that companies that are involved in personalized medicine have biomarkers tied to 73% of their oncology compounds across all phases [38]. The FDA has also approved 15 cancer drugs with companion diagnostics, a number that the Tufts researchers expect to increase markedly in the future, with 50–60% of new cancer medications launched in the next 5 years expected to have a companion diagnostic [38].

According to executives interviewed for the Tufts report, the average company is expected to increase its spending on personalized medicine by 33% over the next 5 years [39]. This investment is fueling a number of deals between diagnostics, biopharmaceutical, and bioinformatics companies. In November 2015, the company Thermo Fisher entered into deals with both Novartis and Pfizer for the development and commercialization of multi-marker, NGS-based companion diagnostic tests that will be directed toward the companies' NSCLC portfolios [40].

Deals have also taken place between Exosome Diagnostics and Eli Lilly for blood-based biomarker discovery, Merck & Co. and Luminex for an Alzheimer's companion diagnostic, and Biodesix and Bruker for mass spectrometry diagnostics [39, 41, 42]. In addition, an entire new type of company is emerging to help move precision medicine ahead. Companies such as DNAnexus (cloud-based bioinformatics) and Cypher Genomics (automated genome interpretation of NGS data for biomarker development) are helping to advance the field by offering advanced genomic annotation services and data integration platforms [43, 44].

In his State of the Union address in January 2015, former President Obama announced a new Precision Medicine Initiative. This would see $215 million of government spending set aside in the 2016 budget for the NIH, the National Cancer Institute (NCI), the FDA, and the Office of the National Coordinator (ONC). This would include $130 million to the NIH for a program to gather the genetic data of more than one million volunteers, $70 million to the NCI to identify the genomic drivers in cancer, $10 million to the FDA to build out an infrastructure that can handle NGS data, and $5 million to the ONC to facilitate clinical data interoperability [45].

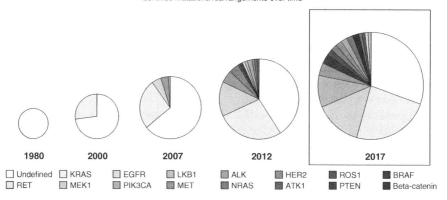

Evolution of lung adenocarcinoma molecular characterization
Identified mutations/rearrangements over time

| 1980 | 2000 | 2007 | 2012 | 2017 |

☐ Undefined ☐ KRAS ☐ EGFR ☐ LKB1 ☐ ALK ☐ HER2 ☐ ROS1 ☐ BRAF
☐ RET ☐ MEK1 ☐ PIK3CA ☐ MET ☐ NRAS ☐ ATK1 ☐ PTEN ☐ Beta-catenin

Figure 1.10 The evolution of lung adenocarcinoma, showing the explosion in the number of molecular targets for therapeutic intervention [36].

Proving that the area of precision medicine is a global movement, there are also a variety of initiatives from governments around the world, including funding from the European Union for programs such as NGS-PTL, an NGS project for identifying prognostic biomarkers in leukemia, and RiskyCAD, which focuses efforts on biomarker discovery and precision medicine development in asymptomatic patients at high risk of coronary disease [25].

The Chinese government also announced its intention to begin integrating personalized medicine into clinical trials in the country. China held its first personalized medicine conference in March 2015, with the Ministry of Science and Technology pledging around $9.5 billion to the area before 2030 [46].

These initiatives clearly signal that precision medicine is now a political as well as a scientific issue on a global scale. And it's not only governments that are putting significant funding into personalized medicine. In November 2015, the Kraft Family Foundation pledged $20 million to the Harvard Business School for the advancement of targeted therapies. Harvard will work with the Broad Institute and other stakeholders in Boston including investors, researchers, and clinicians to fund the development of personalized treatments [47].

Perhaps, more than anything, this move by the Kraft family highlights the challenges facing personalized medicine today. Robert Kraft learned about precision medicine when his wife went through several rounds of chemotherapy for ovarian cancer in 2011, a battle she sadly lost. He felt that despite the incredible promise of targeted therapy, too many stakeholders were working in silos.

The press release from Harvard Business School announcing the donation summed up this problem when it stated that "the growth of the industry (precision medicine) is hampered by gaps that exist between scientific discoveries and the development and commercialization of medical solutions for public benefit. The rising cost of clinical trials and a lack of collaboration among scientists, the pharmaceutical industry and investors also hinder growth" [48].

Building on the initial successes that we have had in precision medicine will require a coordinated effort from all stakeholders—biomedical researchers, public and private investors, policy makers, payers, the biopharmaceutical industry, and individual patients who are willing to give their genetic data in order to advance the field. Additionally, there must be greater push for education on precision medicine as all of these stakeholders' points of view are from different places. Nan Doyle, director of Strategic Partnerships and Development in Pathology at Massachusetts General Hospital says, "I've worked around precision medicine for over 20 years. Education is always the key! Aside from the science ... we will have arrived when every stakeholder understands what precision medicine can do, and what it can't (at least not yet), how to think about health in terms of prevention, and how to be comfortable with a lot of uncertainty while we figure it out. Why? Because we need sustained, foundational support, and we need hope. One way to do that is to increase the throughput in Masters-level genetic counseling training programs. The degree is a hybrid of deep-dive medical genetics and how to talk about it with every possible audience: a skill set that has enormous value outside the clinic. It is truly the MBA of precision medicine, and needs to be just as ubiquitous."

If this disparate group of stakeholders can continue to push and educate each other either formally or informally with an MBA in Precision Medicine, then, in the future, success stories like Caleb Nolan's, our real CF sufferer at the beginning of this chapter, should become more common.

References

1 McMullan D. What is personalized medicine? [Internet]. Genome; [cited Aug 23, 2016]. Available from: http://genomemag.com/what-is-personalized-medicine/#.VkDXBk3lvIU

2 What is cystic fibrosis? [Internet]. Cystic Fibrosis Foundation; [cited Aug 23, 2016]. Available from: https://www.cff.org/What-is-CF/About-Cystic-Fibrosis/

3 Adapted from: Pothier KC, Woosley R, Fish A, Sathiamoorthy T, et al. Introduction to molecular diagnostics [Internet]. DxInsights/AdvaMedDx; 2013 [cited Aug 23, 2016]. Available from: https://dx.advamed.org/sites/dx.advamed.org/files/resource/advameddx_dxinsights_pdf.pdf

4 Pothier KC, Woosley R, Fish A, Sathiamoorthy T, et al. Introduction to molecular diagnostics [Internet]. DxInsights/AdvaMedDx; 2013 [cited Aug 23, 2016]. Available from: https://dx.advamed.org/sites/dx.advamed.org/files/resource/advameddx_dxinsights_pdf.pdf

5 Adapted from: Encyclopædia Britannica [Internet]; 2007 [cited Aug 23, 2016]. Available from: https://www.britannica.com/event/Human-Genome-Project

6 DNA is constantly changing through the process of mutation [Internet]. Nature; 2014 [cited Aug 23, 2016]. Available from: http://www.nature.com/scitable/topicpage/dna-is-constantly-changing-through-the-process-6524898

7 Adapted from: Kalydeco patient education materials [Internet]. Kalydeco; [cited Aug 23, 2016]. Available from: http://www.kalydeco.com/how-kalydeco-works/treating-a-cftr-protein-defect

8 Kovacs D, Gonda X, Petschner P, Edes A, Eszlari N, Bagdy G, et al. Antidepressant treatment response is modulated by genetic and environmental factors and their interactions. Ann Gen Psychiatry [Internet]; 2014 [cited Aug 23, 2016];13:17. Available from: https://www.ncbi.nlm.nih.gov/pmc/articles/PMC4106212/

9 Schork NJ. Personalized medicine: time for one-person trials [Internet]. Nature; Apr 29, 2015 [cited Aug 23, 2016]. Available from: http://www.nature.com/news/personalized-medicine-time-for-one-person-trials-1.17411#/imprecision

10 Franchini M, Mengoli C, Cruciani M, Bonfanti C, Mannucci PM. Effects on bleeding complications of pharmacogenetic testing for initial dosing of vitamin K antagonists: a systematic review and meta-analysis. J Thromb Haemost. Jul 2014;12(9):1480–7.

11 Hitti M. FDA lets MS drug Tysabri return [Internet]. WebMD; Jun 6, 2006 [cited Aug 23, 2016]. Available from: http://www.webmd.com/multiple-sclerosis/news/20060605/fda-lets-ms-drug-tysabri-return

12 Egnew TR. Suffering, meaning, and healing: challenges of contemporary medicine [Internet]. Ann Fam Med Mar 2009 [cited Aug 23, 2016];7(2):170–5. Available from: https://www.ncbi.nlm.nih.gov/pmc/articles/PMC2653974/

13 Pray LA. Discovery of DNA structure and function: Watson and Crick. Nat Educ [Internet]; 2008 [cited Aug 23, 2016];1(1):100. Available from: http://www.nature.com/scitable/topicpage/discovery-of-dna-structure-and-function-watson-397

14 Adapted from: What is protein synthesis [Internet]. Protein Synthesis; Feb 11, 2013 [cited Aug 23, 2016]. Available from: http://www.proteinsynthesis.org/what-is-protein-synthesis/

15 Fredholm L. How the code was cracked [Internet]. Nobelprize.org; Jul 7, 2004 [cited Aug 23, 2016]. Available from: http://www.nobelprize.org/educational/medicine/gene-code/history.html

16 Alberts B, Johnson A, Lewis J, Raff M, Roberts K, Walter P. Molecular biology of the cell. 4th ed. New York: Garland Science; 2002.

17 Saiki RK, Gelfand DH, Stoffel S, Scharf SJ, Higuchi R, Horn GT, et al. Primer-directed enzymatic amplification of DNA with a thermostable DNA polymerase. Science. Jan 29, 1988;239(4839):487–91.

18 What is PCR? [Internet]. Roche Molecular Diagnostics; [cited Aug 23, 2016]. Available from: https://molecular.roche.com/innovation/pcr/what-is-pcr/

19 Porterfield A. The key to genome sequencing came from geysers [Internet]. UA Magazine; Feb 2013 15 [cited Aug 23, 2016]. Available from http://www. ua-magazine.com/the-key-to-genome-sequencing-came-from-geysers/

20 Sinsheimer RL. To reveal the genomes. Am J Hum Genet Aug 2006;79(2):194–6.

21 Hutchison CA. DNA sequencing: bench to bedside and beyond. Nucleic Acids Res. Sep 2007;35(18):6227–37.

22 History of genome sequencing – the Sanger method [Internet]. 454 Life Sciences; [cited Aug 23, 2016]. Available from: https://www.scribd.com/ document/53971660/History-of-Genome-Sequencing-FINAL

23 Bumgarner R. DNA microarrays: types, applications and their future. Curr Protoc Mol Biol. 2013 Jan;0 22:Unit 22.1.

24 Pucheril D, Sharma S. The history and future of personalized medicine. Managed Care [Internet]; Aug 2011 [cited Aug 23, 2016]. Available from: https://www.managedcaremag.com/archives/2011/8/ history-and-future-personalized-medicine

25 Kumar GL, Badve SS. Milestones in the discovery of HER2 proto-oncogene and trastuzumab (Herceptin™) [Internet]. Dako; 2008 [cited Aug 23, 2016]. Available from: http://www.dako.com/28827_2008_conn12_milestones_ discovery_her2_proto-oncogene_and_trastuzumab_kumar_and_badve.pdf

26 Personalized medicine, the right treatment for the right person at the right time [Internet]. European Parliament Briefing; Oct 2015 [cited Aug 23, 2016]. Available from: http://www.europarl.europa.eu/RegData/etudes/ BRIE/2015/569009/EPRS_BRI(2015)569009_EN.pdf

27 Adapted from: Nichols CA. Why elephants almost never get cancer–and why that might save human lives [Internet]. Genetic Literacy Project; Nov 29, 2015 [cited Aug 23, 2016]. Available from: https://www.geneticliteracyproject. org/2015/11/29/elephants-almost-never-get-cancer-might-save-human-lives

28 Biomarkers in risk assessment: validity and validation [Internet]. Inchem.org; 2001 [cited Aug 23, 2016]. Available from: http://www.inchem.org/ documents/ehc/ehc/ehc222.htm

29 Herceptin (Trastuzumab) development timeline [Internet]. Genentech; [cited Aug 23, 2016]. Available from: http://www.gene.com/media/product- information/herceptin-development-timeline

30 Perez EA, Romond EH, Suman VJ, Jeong JH, Sledge G, Geyer Jr CE, et al. Trastuzumab plus adjuvant chemotherapy for human epidermal growth factor receptor 2–positive breast cancer: planned joint analysis of overall survival from NSABP B-31 and NCCTG N9831. J Clin Oncol. Nov 20, 2014;32(33):3744–52.

31 BRCA1 and BRCA2: cancer risk and genetic testing [Internet]. National Cancer Institute; Apr 1, 2015 [cited Aug 23, 21016]. Available from: http:// www.cancer.gov/about-cancer/causes-prevention/genetics/brca-fact-sheet

32 Shen B. Bioinformatics for Diagnosis, Prognosis and Treatment of Complex Diseases. Dordrecht: Springer; 2013.

33 McDermott JE, Wang J, Mitchell H, Webb-Robertson BJ, Hafen R, Ramey J, et al. Challenges in biomarker discovery: combining expert insights with statistical analysis of complex omics data. Expert Opin Med Diagn. Jan 2013;7(1):37–51.

34 Lee D. Next-generation sequencing for disease biomarkers [Internet]. American Laboratory; Dec 15, 2014 [cited Aug 23, 2016]. Available from: http://www.americanlaboratory.com/914-Application-Notes/169217-Next-Generation-Sequencing-for-Disease-Biomarkers/

35 More than 20 percent of the novel new drugs approved by FDA's Center for Drug Evaluation and Research in 2014 are personalized medicines [Internet]. Personalized Medicines Coalition; 2014 [cited Aug 23, 2016]. Available from: http://www.personalizedmedicinecoalition.org/Userfiles/PMC-Corporate/file/2014-fda-approvals-personalized-medicine2.pdf

36 Pothier K, Gustavsen G. Combatting complexity: partnerships in personalized medicine. Pers Med. Jun 2013;10(4):387–96.

37 Woodcock J. FDA continues to lead in precision medicine [Internet]. FDA; Mar 23, 2015 [cited Aug 23, 2016]. Available from: http://blogs.fda.gov/fdavoice/index.php/tag/personalized-medicine/

38 Table of pharmacogenomic biomarkers in drug labeling [Internet]. FDA; Jul 11, 2016 [cited Aug 23, 2016]. Available from: http://www.fda.gov/Drugs/ScienceResearch/ResearchAreas/Pharmacogenetics/ucm083378.htm

39 Merck and Luminex Corporation enter agreement to develop companion diagnostic to support investigational BACE inhibitor clinical development program for Alzheimer's disease [Internet]. Merck; Mar 13, 2013 [cited Aug 23, 2016]. Available from: http://www.mercknewsroom.com/press-release/research-and-development-news/merck-and-luminex-corporation-enter-agreement-develop-co

40 Shaffer AT. Personalizing cancer drugs: the next front in diagnostics [Internet]. OncLive; Oct 28, 2015 [cited Aug 23, 2016]. Available from: http://www.onclive.com/publications/oncology-live/2015/october-2015/personalizing-cancer-drugs-the-next-front-in-diagnostics/1

41 Thermo Fisher inks companion Dx deal with Novartis, Pfizer [Internet]. GenomeWeb; Nov 18, 2015 [cited Aug 23, 2016]. Available from: https://www.genomeweb.com/sequencing-technology/thermo-fisher-inks-companion-dx-deal-novartis-pfizer

42 Exosome diagnostics enters collaboration agreement with Lilly for exosome blood-based biomarker discovery [Internet]. Exosomedx; Sep 19, 2013 [cited Aug 23, 2016]. Available from: http://www.prnewswire.com/news-releases/exosome-diagnostics-enters-collaboration-agreement-with-lilly-for-exosome-blood-based-biomarker-discovery-224386351.html

43 Biodesix inks deal with Bruker for mass spec support; closes on $8.8M in Series D financing [Internet]. GenomeWeb; Apr 11, 2013 [cited Aug 23, 2016]. Available from: https://www.genomeweb.com/proteomics/biodesix-inks-deal-bruker-mass-spec-support-closes-88m-series-d-financing

44 DNAnexus [Internet]; [cited Aug 23, 2016]. Available from: https://www.dnanexus.com

45 Cypher genomics [Internet]; [cited Aug 23, 2016]. Available from: http://cyphergenomics.com

46 Quinn B. White House details $215M initiative in precision medicine [Internet]. Discoveries in Health Policy; Jan 30, 2015 [cited Aug 23, 2016]. Available from: http://www.discoveriesinhealthpolicy.com/2015/01/white-house-details-215m-initiative-in.html

47 Salvacion M. China to begin R&D on precision medicine [Internet]. Yibada; Nov 2, 2015 [cited Aug 23, 2016]. Available from: http://en.yibada.com/articles/80433/20151102/china-begin-r-d-precision-medicine.htm

48 Robert and Myra Kraft Family Foundation establish $20 million endowment at Harvard Business School [Internet]. Harvard Business School; Nov 18, 2015 [cited Aug 23, 2016]. Available from: http://www.hbs.edu/news/releases/Pages/kraft-family-foundation-establishes-endowment.aspx

2

Decision-Making Machines

Diagnostics in Precision Medicine

David Hungerford was 32 years old the first time he saw white blood cells from a patient with leukemia. The disease was chronic myeloid leukemia (CML), so these cells were both the disease's sign and its instrument of destruction. In a healthy body, these particular types of white blood cells are among the first responders to infection, or injury, or an attack of parasites; in the patient where Hungerford's sample originated, as in all CML patients, they would multiply out of control, taking over the bone marrow (also called myeloid tissue) where they are born and mature. The year was 1959, and the causes of CML—or any cancer—were a mystery at that time.

The cells had come to him by way of Peter Nowell, a pathologist at the University of Pennsylvania. Nowell had grown them in a culture in his lab, watching them replicate, studying their growth and proliferation, and searching for any weakness that he or others could exploit. As Nowell's investigation wound down, he "decided it was a shame to throw the samples away," as he said years later, and he shared them with Hungerford [1]. Nowell had had little, if any, idea what he was looking for. Neither did Hungerford, who, as an early graduate student, was focused on that most fundamental of scientific problems: finding a subject to write his dissertation on.

As Hungerford examined the cells under a light microscope—technologically the same instrument I used in my high school biology class—he focused on its chromosomes, the 23 pairs of structures (one-half from each parent) that carry nearly all our DNA.[1] In 1959, scientists had only just figured out how many chromosomes there were; the structure of DNA itself had been published just 6 years earlier, in a scant single page of the journal *Nature*. By using novel techniques to stain, or dye, and visualize the chromosomes, Hungerford

---

1 A small amount of DNA is also found in a different cell structure, the mitochondrion, which is inherited entirely from the mother.

Personalizing Precision Medicine: A Global Voyage from Vision to Reality, First Edition.
Kristin Ciriello Pothier.
© 2017 John Wiley & Sons, Inc. Published 2017 by John Wiley & Sons, Inc.

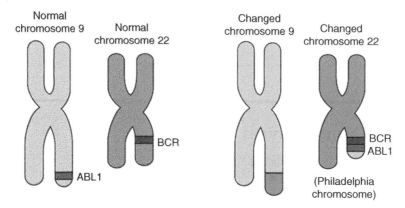

Figure 2.1 Changed chromosomes 9 and 22 [2].

noticed something odd: in the leukemia patient's cells, part of chromosome 22 was missing (Figure 2.1) [3]. In fact, half of it was gone [2].

Further work would clarify Hungerford's findings, scrubbing them of errors and bringing them into higher resolution. For instance, Hungerford had mistakenly identified the chromosome *manqué* as number 21, not 22. Moreover, the missing piece wasn't truly gone: it was eventually found, instead, attached to chromosome 9. Number 9, in turn, had lost part of its end to chromosome 22—a switch that, in the field, is called a translocation. Hungerford's discovery, though, made him the first person to find decisive evidence that cancer—for millennia, an awful mystery—was fundamentally a genetic disease. It would also mark the first of many firsts for CML. The disease not only taught scientists of the 1950s a basic truth about cancer but also became one of the earliest triumphs, if not the very earliest, for precision medicine.

Fourteen years after Dr. Hungerford—who did, in time, finish his Ph.D.— identified the broken chromosome, Dr Janet Rowley discovered an important limitation of the 9–22 translocation [4]. Most genetic diseases are, by their nature, heritable. A mutation exists in a person's DNA from conception, from the time that person is a single cell, and is perpetuated in every cell of the body as that single cell divides countless times to become a fully formed human. Since "every cell," in this case, includes eggs and sperm, the mutation would also be passed onto the person's children. But the translocation—what came to be called the Philadelphia chromosome, in honor of Hungerford's and Nowell's labs at the University of Pennsylvania in Philadelphia—isn't like that. This mutation happens spontaneously in the rapidly dividing cells of the bone marrow, where a stash of progenitor cells, cells destined to differentiate into specific kinds of cells, is constantly changed into specialized blood cells, red or white, in response to various signals from the body. In nascent white blood cells, and apparently only there, the Philadelphia chromosome can appear.

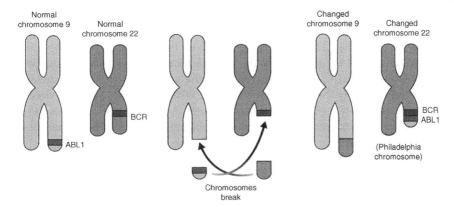

Figure 2.2 ABL and BCR gene combination [3].

At first, it wasn't clear why the mutant chromosome would cause the wild, dysregulated growth of leukocytes that characterizes leukemia. (The suffix *-emia* refers to something found in the blood, so the word *leukemia* just refers to leukocytes in the blood; that's where they're supposed to be, but the term implies an excess of them.) That discovery came 9 years later, in 1982, when researchers found the specific genes that were carried in the translocated parts of chromosomes 9 and 22. When *ABL*, the gene on 9, combines with *BCR*, on 22, the result was a new hybrid gene—which produced a new hybrid protein (Figure 2.2) [3, 5].

The hybrid is disastrous, almost perfectly engineered for mayhem. ABL, by itself, is a type of protein called a tyrosine kinase, which is crucial to most of what nearly all the cells in the body do. Picture each cell of your body as a full, absolutely packed set of machinery, with an intensely complex set of balances and counterbalances all happening at once. Is the cell going to divide? Does it need to stop dividing? Does it need to increase its metabolism, or decrease it, or make more or less of any of the countless proteins it could, conceivably, make? Most cells can respond in any of these ways to an equally countless number of stimuli, including signals (themselves proteins made by other cells in response to other signals) from elsewhere in the body. To be so flexible, a dazzlingly intricate machinery is constantly in motion, being accelerated or decelerated, balanced or counterbalanced. Tyrosine kinases are used inside each cell to control that machinery. At any given second, in your own body, tyrosine kinases do their job somewhere in the neighborhood of millions to billions of times.

But the tyrosine kinases, themselves, need to be controlled. If they act as an accelerator pedal, then there needs to be a sensitive foot on the pedal, modulating how intensely it acts. The foot, in this case, is also part of the signaling network.

ABL, like most tyrosine kinases, has a range of jobs, and, outside the context of the Philadelphia chromosome, it's perfectly controllable. BCR, though, changes everything. The BCR–ABL hybrid, or fusion protein, can't be controlled. The pedal is constantly to the floor. A number of processes in the cell, all controlled by ABL, are accelerated, with the net result that the cell divides much more quickly than it otherwise would. Worse, the cell's ability to detect and fix errors as it replicates is much reduced. As cells replicate themselves, copying their DNA over, mistakes are inevitable, even constant, but a multiply-redundant system is in place to stop the assembly line. If, in the direst of circumstances for the cell, an error can't be fixed, the cell self-destructs to prevent damaged DNA from being loosed into the body. (The process is called *apoptosis*; the second P is silent, like in "pterodactyl.") BCR–ABL defeats this entire quality-control mechanism, and damaged, poorly developed leukocytes flood the bone marrow (Figure 2.3) [6]. This is CML.

In a patient, the first sign of the disease is usually the unusual number of leukocytes present in the blood. At that point, nothing may even have seemed wrong, as far as that person could tell. What confirms the diagnosis is a process very much like what Hungerford did—staining chromosomes and studying them under a microscope, called a cytogenetic test. (The sample would be taken in a biopsy of the bone marrow.)

Through the 1990s and as recently as the early 2000s, this was the only diagnosis anyone needed, since the options for treatment were so feeble and so poor (Figure 2.4) [7]. Patients could receive cytotoxic chemotherapy, a series of drugs meant essentially to poison dividing cells; since the dysregulated cells of leukemia divide so much more quickly than any normal cell, they would be disproportionately affected—but the collateral damage across the body is severe. The intense nausea and vomiting associated with chemotherapy, both in practice and in the popular imagination, comes largely from this cytotoxic kind. Another option, a protein called interferon alpha, would rally the body's

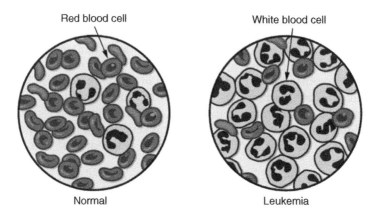

Figure 2.3 Abnormal white blood cell counts in a leukemia patient [6].

own immune system to its defense, but imprecisely; patients generally feel like they have a relentless, severe flu. Neither cytotoxic chemotherapy nor interferon alpha provided a cure; the best they could do was to temporarily deplete the number of leukemia cells before their inevitable resurgence. On the other hand, a bone marrow transplant—destroying every last cell in the body's bone marrow and replacing it with marrow from a donor—was a cure, but a risky, difficult one [8, 9]. All in all, even with extraordinary and heroic measures, a diagnosis of CML gave patients an average of 5 years to live. The situation was painful for patients—and frustrating for the doctors who cared for them. "Chronic myeloid leukemia is an enigma in origin and a source of frustration in treatment," the legendary oncologist Richard T. Silver wrote [10].

Paul, a 34-year-old psychology Ph.D. who taught at a medical school, received his diagnosis in the course of a single day in 2006. His doctor called to alert him to his white blood cell count, which was alarmingly high, and urge him to go to the hospital immediately, with no time to lose. Paul called his parents to watch his children. Then he was in his car, and then the emergency room, and then, that evening, a member of the oncology team on call was telling him that he had CML [11].

At that moment, Paul had an advantage that no one in history, until a few years earlier, had had—a drug called Gleevec. In 1993, a young oncologist in Portland named Brian Druker had decided he couldn't stomach the career he saw ahead of him, telling patients with CML that they had 5 years or less to live, and set out to develop a drug to attack and suppress BCR–ABL. In 1996 he had a contender,

Chemotheraeutic drugs used to treat the chronic phase of chronic myeloid leukemia

Drug	Dose[a]	Adverse effect[b]
Hydroxyurea	0.5–2.0 g/day orally	Cytopenias, rash, nausea
Busulfan	0.5–2.0 g/day orally	Cytopenias, rash, bone marrow aplasia
Interferon alfa	5 million U/m²/day subcutaneously	Fever, myalgias, rash, depression, thrombocytopenia
Interferon alfa plus cytarabine	Interferon alfa, 5 million U/m²/day subcutaneously 20 mg/m²/day for 10 days each month	Fever, myalgias, rash, depression, thrombocytopenia, nausea, vomiting, diarrhea, mucositis, weight loss

[a]Doses are modified on an individual basis according to changes in the patient's peripheral-blood counts.

[b]Data are from randomized clinical trials.

Figure 2.4 Comparison of chemotherapeutic drugs [7].

which the newly formed Swiss drug company Novartis launched into clinical trials in patients. Known, then, by its generic name, imatinib, Gleevec penetrates myeloid cells in the bone marrow and latches onto active BCR–ABL, blocking it from causing havoc inside the cell. Shielded from BCR–ABL, cells return to dividing at normal rates, repairing flaws in their DNA and healthily self-destructing via apoptosis when needed. To patients, the effect is clear. Gleevec "produces complete hematologic and cytogenetic responses in a substantial percentage of chronic myeloid leukemia patients" [12]. Patients who, once, would have been given 5 years of survival started living out normal lives. They faced an average life expectancy of 72 years for men and 78 for women [13]—essentially the same as those who never had leukemia at all. And all this thanks to a drug that was taken once a day, by mouth, at home and caused only mild side effects. Dr Harold Varmus, then the president of Memorial Sloan Kettering Cancer Center—not a position that tolerates hyperbole—was quoted (in *People* magazine, of all places) calling it a "huge triumph" [14].

Paul started his own treatment with Gleevec shortly after his diagnosis. Over the course of the next year, he found himself undergoing an enormous range of diagnostic tests—the entire armamentarium for understanding CML, developed over the decades that followed David Hungerford's first view through his microscope. Observing Paul's path through treatment, it becomes increasingly clear that the term "diagnostic" for many of these tests is a misnomer, or at least incomplete. Paul's physicians used laboratory testing not just to *diagnose* his disease—the initial blood cell count and cytogenetic test took care of that—but also to monitor his own individual response to his treatment and adjust the therapy accordingly. CML is one of the first diseases for which medicine became, legitimately, personalized and precise.

From the beginning, the monitoring tests that Paul went through showed that he was on track. The first milestone Paul needed to reach was called a *hematologic response*—a sign that the white cell count in his blood had stabilized. Within 2 weeks, the breakneck growth of his leukemia had not just slowed, but stopped altogether. At the 6-month mark, when Paul had his first bone marrow biopsy since diagnosis, he had reached the second milestone, a *complete cytogenetic response* (CCyR). The term "complete" here is a little misleading but has a precise definition: cytogenetic testing of his bone marrow found that the population of Philadelphia chromosomes, among all the cells in his marrow, had decreased by 99%.

Ninety-nine percent constitutes a strong performance in most spheres of life. But for leukemia, where the number of corrupted cells stretches deep into the millions, 99% of the way there is just getting started. To measure the response to Gleevec any more precisely requires a more exacting toolkit—in this case, an amplifying technology called PCR (see Chapter 1).

By using PCR, a lab can amplify vanishingly tiny amounts of DNA into a quantity that's actually realistic to measure. In the case of CML, the genes targeted for amplification are *BCR–ABL* and a second gene, known as a reference gene. The reference gene is usually one that is mutually exclusive with BCR–ABL—that is, it wouldn't be present if *BCR–ABL* were. Typically, *BCR* or *ABL* by themselves, in unfused form, are used as reference genes. After the amplification step, other techniques are used to quantify the amount of each gene, and the quantity of *BCR–ABL* is reported relative to the level of the reference gene. In the 2000s, as Paul was starting his treatment, the most sophisticated labs were able to detect a 99.9%, or 1000-fold, reduction in *BCR–ABL* by PCR. In the language of CML treatment, this would be an MR3—MR for *molecular response* and 3 for the reduction on the logarithmic scale (where a 1-log reduction would be 10-fold, 2-log 100-fold, etc.).

At 6 months, corresponding with his 99% cytogenetic response, Paul's molecular response was 2-log, or 100-fold. His target, though, was more ambitious. Patients who reach MR3 are said to have achieved a *major molecular response*. What made this, for Paul, the most meaningful milestone of all is that patients who reach it in 12–18 months, who are on a kind of therapeutic fast track, fare much better in the long run than those who don't. They are less likely to have a severe recrudescence of leukemic cells, the intense and difficult-to-treat flare-ups called blast crises, and this lower likelihood translates into a longer, sounder life expectancy. Years of clinical data had established and validated these milestones on a global scale. In fact, they're so critical that researchers ensured that MR3 has precisely the same meaning everywhere. For Paul, it isn't a 1000-fold reduction in *BCR–ABL* relative to *his own* initial levels; instead, the initial reference point is an internationally agreed-upon sample. Working shortly after Gleevec's launch in 2002, a team in Australia gathered 30 patients who hadn't yet been treated, averaged their *BCR–ABL* levels, and set that as, effectively, MR0. To this day, an MR3.0 response is a 1000-fold reduction relative to those 30 Australians, and everyone at MR3.0 has (more or less) exactly the same amount of *BCR–ABL* DNA circulating in his or her marrow.

Eighteen months after he first started Gleevec, just in time to stay on the fast track, Paul managed to reach MR3. He isn't cured, though; no one is with CML. He has a lifelong disease and a treatment to take every day for the rest of his life to keep the leukemic cells suppressed. And that's the best-case scenario. Patients can eventually develop resistance to Gleevec; in fact, as many as 3 in 10 do.

For these patients, the result is similar to what would have happened to Paul had he missed any milestones along the way: their treatment changes. The dose of Gleevec could be increased; alternately, the patient could be switched to a different drug altogether. In the mid-2000s, the world of CML had come a long way from metabolic poisons and annihilating bone marrow. Two other treatments aimed at BCR–ABL had followed—Sprycel (or dasatinib) in 2006 and Tasigna (or nilotinib) in 2007—and, together, the three drugs came to be known as a class: the tyrosine kinase inhibitors (TKIs).

Today, a patient could start treatment on any of the three and, if treatment doesn't stay on track, switch to another. (Two other TKIs—Bosulif, or bosutinib, and Iclusig, or ponatinib—are available to patients who have failed multiple TKIs.) Treatment for CML, in other words, is precision medicine—rather than having every CML patient follow the same script, patients and their physicians can pivot quickly away from treatments that don't seem to do the job.

As therapeutics have become more sophisticated, so, too, have the diagnostic and monitoring tools used to guide them. The most recent widely available diagnostics are able to detect an MR4.5—a reduction in *BCR–ABL* of 32,000-fold. Several studies, still ongoing, suggest that patients who reach MR4.5 and sustain it over time may have so thoroughly suppressed BCR–ABL that they can take a holiday from treatment, of indefinite duration. This level of response, called treatment-free remission (TFR), still can't rightly be called a cure, but it sets a new standard for therapy in CML.

It's clear, though, that nothing about this ideal can work if patients' *BCR–ABL* levels aren't monitored—regularly and frequently. Medical associations in both the United Stated and the EU have published guidelines that recommend the optimal tests and intervals for the first 4 years of treatment [15]. In these recommendations—compiled by working groups of top oncologists—patients rarely, if ever, go more than 3 months without a check-in. (The US guidelines recommend, in total, roughly 14 *BCR–ABL* PCR tests over 4 years; the European guidelines call for 11.) The evidence to support this approach is fairly stark: if one group of patients is tested three to four times a year and another group isn't tested at all, then, by the end of 3 years, eight times as many people in the second group will have sickened or died [15].

Of course, the desire to be tested or to test one's patients is not, by itself, enough. In some countries or regions, a lab that's capable of carrying out an accurate *BCR–ABL* may be hundreds of miles away–across deserts, over seas, or through trackless forests. Even if the lab is accessible, the test itself may not be: many *BCR–ABL* tests cost in the vicinity of $100–300 per sample, and even some rapidly developing markets (notably Brazil and Russia) leave patients without private insurance almost completely on their own to pay the bill. As a result, CML highlights a struggle of precision medicine as much as it highlights its successes: no patient can benefit from a test he/she doesn't receive.

Still, CML, after the birth of the TKIs, remains one of the few conditions that prompt sober, cautious oncologists, *en masse*, to use words like "miracle." Paul, for instance, recently celebrated his 10th year of MR3. CML is unquestionably one of the great success stories of precision medicine. David Hungerford never got to see it: he died, of cancer but not of leukemia, in 1993. That was the same year that Brian Druker, incensed at the powerlessness of modern medicine, opened his lab at Oregon. Even today, we can only aspire to have a Gleevec for every cancer and a test for every Gleevec.

References

1 Altman LK, David A. Hungerford dies at 66; found genetic change in cancer [Internet]. New York Times; Nov 5, 1993 [cited Dec 7, 2016]. Available from: http://www.nytimes.com/1993/11/05/obituaries/david-a-hungerford-dies-at-66-found-genetic-change-in-cancer.html

2 Adapted from: Chronic myelogenous leukemia treatment [Internet]. National Cancer Institute; [cited Dec 7, 2016]. Available from: http://www.cancer.gov/types/leukemia/patient/cml-treatment-pdq

3 History [Internet]. Philadelphia Chromosome; [cited Dec 7, 2016]. Available from: http://pubweb.fccc.edu/philadelphiachromosome/history.html

4 Rowley JD. A new consistent chromosomal abnormality in chronic myelogenous leukaemia identified by quinacrine fluorescence and giemsa staining. Nature. Jun 1, 1973;243(5405):290–3.

5 de Klein A, van Kessel AG, Grosveld G, Bartram CR, Hagemeijer A, Bootsma JR, et al. A cellular oncogene is translocated to the Philadelphia chromosome in chronic myelocytic leukaemia. Nature. Dec 23, 1982;300(5894):765–7.

6 What does a low white blood cell count indicate? wiseGEEK; [cited Dec 7, 2016]. Available from: http://www.wisegeekhealth.com/what-does-a-low-white-blood-cell-count-indicate.htm

7 Adapted from: Sawyers CL. Chronic myeloid leukemia. N Engl J Med. April 29, 1999;340(17):1330–40.

8 Negrin RS, Schiffer CA. Overview of the treatment of chronic myeloid leukemia [Internet]. UpToDate; 2015 [updated Nov 4, 2016; cited Dec 7, 2016]. Available from: http://www.uptodate.com/contents/overview-of-the-treatment-of-chronic-myeloid-leukemia

9 Cortes JE, Silver RT, Khoury HJ, Kantarjian HM. Chronic myeloid leukemia [Internet]. Cancer Network; Jun 1, 2016 [cited Dec 7, 2016]. Available from: http://www.cancernetwork.com/cancer-management/CML

10 Silver RT. Chronic myeloid leukemia. Curr Opin Oncol. Feb 1992;4(1):66–72.

11 Gershon J. Jon's CML Diary [Internet]. Jon's CML Diary; 2010–2016 [cited Dec 7, 2016]. Available from: http://jongershon.blogspot.com

12 Druker BJ. Imatinib and chronic myeloid leukemia: validating the promise of molecularly targeted therapy. Eur J Cancer. Sep 2002;38(Suppl 5):S70–6.

13 Verma D, Kantarjian H, Strom SS, Rios MB, Jabbour E, Quintas-Cardama A, et al. Malignancies occurring during therapy with tyrosine kinase inhibitors (TKIs) for chronic myeloid leukemia (CML) and other hematologic malignancies. Blood. Oct 20, 2011;118(16):4353–8.

14 Charles N. The miracle worker. People; Feb 19, 2001.

15 Goldberg SL, Chen L, Guerin A, Macalalad AR, Liu N, Kaminsky M, et al. Association between molecular monitoring and long-term outcomes in chronic myelogenous leukemia patients treated with first line imatinib. Curr Med Res Opin. Sep 2013;29(9):1075–82.

3

Precision Medicine around the World

Europe

A war against cancer is being fought inside an unobtrusive glass building nestled among shops and restaurants 5 miles outside of Paris. Inside these glass walls, the Institut National du Cancer (INCa), also known as the French National Cancer Institute, is coordinating battle plans and deploying resources across its network of 28 platforms for molecular genetics (see Figure 3.2). This well-oiled machine provides free innovative care for everyone in France and has made tremendous strides in cancer research, screening, and treatment.

France is one of several European countries excelling in precision medicine today, with each country in Europe in pseudo competition with one another and jockeying for position (Figure 3.1). The United Kingdom led the consolidation of next-generation sequencing (NGS) offerings into multiple formidable labs some years ago. In Italy, there is access to NGS, but the reimbursement for testing is sometimes in question, which dissuades physicians from ordering, whereas reimbursement is further ahead in France, in which INCa provides funding (albeit temporary) to molecular testing centers and ensures that companion diagnostics are available and funded as soon as a new drug is available. And across Europe (and worldwide), the company Roche/Foundation Medicine is setting up processes for countries to access its offerings of its FoundationOne® genomic testing in addition to major labs offering various forms of testing, providing additional competition, and making the way for each country to reassess its environment with respect to the clinical laboratory.

In France, Italy, Germany, and the United Kingdom, an average of 3.8 oncologists exists for every 100,000 people. There is a large range by country however, with France matching more to the United States (1.3 per 100,000) and Italy one of the highest at 7.4 per 100,000. And Europe provides more oncologists versus many other countries, such as Brazil (1.2) or China (0.6) [2–7].

Personalizing Precision Medicine: A Global Voyage from Vision to Reality, First Edition.
Kristin Ciriello Pothier.

Europe facts

Population (2015): 738 million

GDP (2016): $19 trillion

Cancer incidence (2012): 3.45 million

Map:

Figure 3.1 European demographic facts.

Europe in general has made tremendous progress in precision medicine, not only in scientific advances but also in ensuring that much of its populace has relatively affordable and equitable access to personalized medicine. This chapter will explore precision medicine advances in France, the United Kingdom, and Germany, who are not only the most populated and highest GDP countries in Europe (at the writing of this book, "Brexit," the UK's

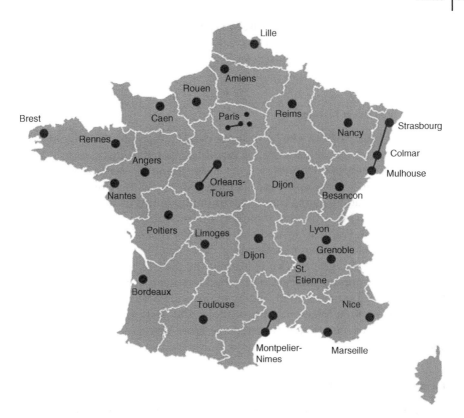

Figure 3.2 Distribution of INCa platforms in France [1].

prospective withdrawal from the European Union, is still in process) but also leaders who have made great strides in the field.

France

France's efforts with screening, biomarker testing, treatment, and continued investment in cancer research has propelled it to the forefront of precision medicine advances.

Initial Cancer Screening Increased

Through persistent screening programs, INCa has increased the breast cancer screening participation rate from 43% in 2004–2005 to 52% in 2014. Those in the target demographic (i.e., 50–74-year-old females) receive invitations for free screenings every 2 years by mail and a second invitation if they do not see their general practitioner within three months of the first invitation [8]. Since 2005, breast cancer mortality rates have decreased by 1.5% each year in France due in large part to early diagnosis as well as advances in targeted therapies and other therapeutic care [9].

Focus on Biomarker Testing

After screening, patients who have cancer may undergo additional biomarker testing, which is free to patients and readily accessible at institutions within INCa's well-equipped laboratories. As more therapies that target various genetic alterations are approved, pathologists must use tiny amounts of tissue from a patient's lung tumor, for example, to test for multiple genetic alterations. Having sufficient amounts of tissue sample, ensuring a speedy turnaround time for testing, and decreasing the cost of testing for multiple genetic alterations remain challenging. NGS has the potential to address all of these challenges. France is leading the pack across Europe when it comes to the implementation and reimbursement of novel technologies like NGS. INCa's success in providing nationwide access to molecular testing and its swift implementation of molecular tests for new tumor biomarkers help to ensure that targeted therapies reach the right patients at the right time.

Increased Treatment Access

Patients in France who test positive for biomarkers for which there are targeted therapies have access to targeted therapies, also at no cost to the patient. For those with an advanced refractory malignancy and no therapeutic alternatives, France provides access to clinical trial drugs through the AcSé program. For instance, although crizotinib is only approved for use in non-small cell lung cancer patients who are ALK positive, the AcSé program has provided more than 100 patients in France with access to crizotinib through clinical trials. Among these patients, 39 have lung cancer with a ROS1 genetic alteration. Although crizotinib is not yet approved in patients with ROS1 genetic alterations, recent data indicates that 72% of the 39 AcSé patients with ROS1 had a tumor response and 44% had no progressive disease after 12 months of treatment. Without the AcSé program, these patients, who had exhausted other options, would not have had access to any therapies to manage their lung cancer.

INCa's efforts have had a significant impact on cancer mortality rates in France. Since INCa's founding in 2004, the mortality rate in France has dropped from the second highest to the second lowest in the EU5 (see Figure 3.3) [10].

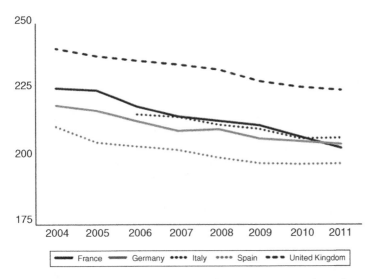

Figure 3.3 Deaths per 100,000 population from malignant neoplasms from 2004 to 2011 (the last year for which data is available from the OECD for all EU5 countries) [10].

To ensure continued success in precision medicine, France has contributed to the development of centers like Genopole, France's leading biotech and bio-therapy cluster. I went for a visit in its facilities just south of Paris, and while there were three train transfers and one very strong espresso to navigate before arriving, it was worth the trip to witness firsthand what has been built. With 81 biotech companies and 20 academic research labs co-located on more than one million square feet of space, Genopole brings together industry, academia, and government institutions to accelerate development in precision medicine. Over a summer lunch and a second espresso, I met with the management team to discuss Genopole's progress on its multiple initiatives. With so much inter-national competition for research grants and talent in Europe and worldwide, the team believes it has to put extra emphasis on supporting INCa and estab-lishing a separate state-of-the-art reference center for innovation, expertise, and knowledge transfer. The biggest challenge in France, the team explained, is educating all of the stakeholders in country who have been treating their patients one way for their entire careers that there are other ways and that technology and treatment are actually accessible in their home country. To meet this need, Genopole is developing this center of excellence they are call-ing CRefIX. CRefIX will be designed to house labs to test innovative platforms, store epidemiological data, and run educational programs for researchers, physicians, and sequencing centers in France. "Genopole, as France's leading biocluster, is entering the age of personalised medicine and will drive the

Reference, Innovation and Expertise Center (CRefIX) to assist in all R&D aspects relating to the development and monitoring of the twelve sequencing units of France Médecine Génomique 2025," says Pierre Tambourin, founding CEO of Genopole.

In addition to bringing individuals together to share information and provide training, another part of the Genopole's value proposition is its ability to provide shared resources. While a start-up may not be able to afford expensive equipment on its own, Genopole offers shared sequencing and imaging platforms to its members. By leveraging Genopole's community and facilities, Genopole members have filed more than 1000 patents since 2009 and have 38 products in development in stages ranging from preclinical to commercial launch [11].

While several countries in Europe compete for the development and execution of the most advanced precision medicine, France's commitment to clinical diagnostics access through the INCa network allows it to stand out just a little bit more. France has created an environment that fast-tracks advances in precision medicine and that provides equal access to healthcare. While cancer and other diseases are still a significant source of mortality in France and elsewhere, France's achievements through INCa, Genopole, and other organizations demonstrate how much can be gained through the effective implementation of precision medicine investment.

United Kingdom

Across the English Channel just 20 miles from France, the United Kingdom is hard at work with a goal of establishing *itself* "as the national and international leader and convener of the precision medicine industry" [12]. The United Kingdom has been at the forefront of genomic discoveries over the last century. It was Watson and Crick in the United Kingdom who won a Nobel Prize for discovering the double-helix structure of DNA. Fred Sanger, a British scientist, then discovered how to sequence DNA. In 2003, British scientists along with many international collaborators had succeeded in sequencing all three billion base pairs of the entire human genome, an achievement that took 13 years and over $3 billion, through the Human Genome Project, mentioned earlier [13].

Now that one human genome has been sequenced, Genomics England in the United Kingdom has begun to sequence 100,000 more genomes through the 100,000 Genomes Project, which was launched in 2012 and is currently the largest national sequencing project of its kind. As the name suggests, the objective of the project is to sequence 100,000 whole genomes by 2017. Fortunately, it no longer costs $3 billion nor takes 13 years to sequence one genome; the cost to sequence a whole genome has decreased to less than $1500 per genome and takes just a few days. With the genomic data from the 100,000 Genomes

Project, scientists can investigate links between genetic data and disease with the end goal of developing new and more effective treatments [13].

The focus of the 100,000 Genomes Project has been on patients with rare diseases and/or cancer because these disorders are so often linked to the genome. At least 80% of rare diseases are believed to be genomic. Given that most rare diseases are inherited, the 100,000 Genomes Project will aim to sequence the genome of a person with a rare disease as well as two of their closest blood relatives to pinpoint the genomic cause of the disease. Cancer is also a focus area given that cancer arises as a result of changes in genes within what were previously normal cells. A person's DNA can develop mutations or changes that enable a tumor to grow and spread, and analyzing these genetic mutations will allow researchers to learn more about cancer, its potential causes, and even potential treatments [13].

The United Kingdom is particularly well suited to lead the 100,000 Genomes Project—due to the UK's provision of universal healthcare through the National Health Service (NHS), it has the ability to obtain and link a lifetime of medical records on a large scale to genomic data from the 100,000 Genomes Project. This is far less feasible in countries with less comprehensive medical records and countries where health records are scattered across multiple public and private insurers.

Benefits from the 100,000 Genomes Project are already being seen. Leslie Hedley, who volunteered to participate in the 100,000 Genomes Project, has a lifelong history of high blood pressure, which has led to kidney failure. Other family members, including his father, brother, and uncle, died of the same condition. Leslie chose to participate in the 100,000 Genomes Project so that he could learn more about his condition in hopes that his daughter, who has early signs of kidney damage, can avoid kidney failure and the health issues that Mr. Hedley has experienced. By sequencing his genome, researchers found that Mr. Hedley's kidney failure was caused by a genetic variant that, fortunately, can be effectively controlled through drugs that are already available. Diagnosis through the 100,000 Genomes Project was able to significantly improve health outcomes not only for Mr. Hedley but also for his daughter Terri and potentially his granddaughter Katie [14].

The 100,000 Genomes Project is one of many precision medicine initiatives in the United Kingdom. The public sector and charitable bodies in the United Kingdom have funded more than 400 infrastructure assets, including 89 research and training centers that are working on precision medicine. Given the high number of entities involved in precision medicine in the United Kingdom, coordination is key. Innovate UK and its Programme Coordination Group serve to identify groups that are conducting similar work in terms of function and disease area, which can serve to minimize overlap and increase collaboration as the United Kingdom continues to build upon its legacy of genomic discovery [15]. Innovate UK has set up organizations called "Catapult centres" that are meant to coordinate the collaboration between scientists, engineers, and market opportunities in order to promote research and development [16]. There are life science-specific Catapults including the Cell and Gene Therapy Center based at Guy's Hospital, London, and the Precision Medicine Catapult based in Cambridge [17]. The Precision Medicine Catapult intends to accelerate precision medicine to become a "mainstream" healthcare solution. It looks to help develop and grow precision medicine more quickly for the benefit of patients and the life sciences industry while driving economic growth for the United Kingdom. Its ultimate goal is to make innovations in precision medicine so that it is readily available for all patients [12]. Another similar initiative that was intended to bridge the so-called valley of death for developmental assets is the Diagnostic Evidence Co-operative (DEC) of London. The DEC is a partnership between Imperial College Healthcare NHS Trust and Imperial College London that is aimed to develop new methods, generate evidence, and integrate *in vitro* diagnostics into clinical practice [18]. DEC has developed a diagnostic research toolkit for efficient generation of evidence of diagnostic tests to ensure the accuracy, usability, and safety of these procedures in order to positively impact patient care [19].

The National Institute for Health and Care Excellence (NICE) and NHS England jointly released a proposal last October to streamline and expedite its technology appraisal process with intention to speed access to innovative products. Under the new proposal, there would be a new "fast-track" NICE technology appraisal process for the most promising new technology in order to get these treatments to the patients more quickly. Although to quality, they would have to fall below an incremental cost-effectiveness ratio of £10,000 per quality-adjusted life year (QALY) [20]. Under this proposed "fast-track" process, technologies would be funded by NHS England within 30 days after NICE publishes its final guidance, much faster than the current 90-day wait time [21].

Additionally, the long-awaited ministerial Accelerated Access Review (AAR) was finally published in late October, which provides some insight on precision medicine. The report goes into detail about the need to streamline new products to market in order to increase market accessibility to patients in need [22]. Among many other things articulated therein is a commitment to form a new Accelerated Access Partnership (AAP) body, which is described to be a "light-touch umbrella organization that brings together the existing activities of National Institute for Health Research, Medicines and Healthcare Products Regulatory Agency, NICE, NHS England, the Department of Health, and NHS Improvement." This partnership will align each of the agency's innovation-related functions around the mission of accelerating patient access to key products [22]. The AAP will look to scan and prioritize the strategically important products, articulate the healthcare system's priorities to innovators, and help focus transformative medicine and design through investment [22]. This new

Did You Know?
The European Personalised Medicine Association (EPEMED) is a nonprofit founded in 2009 by a group of European leaders with background and expertise in the application and development of diagnostic tools and stratified medicine. It is an independent association that focuses exclusively on personalized medicine industry issues within Europe [23]. EPEMED interacts with regulators, payers, and government in order to drive adoption of and access to personalized medicine throughout the European continent. Their mission is to advance personalized medicine in Europe and act as an independent voice and a catalyst to promote the role of diagnostics and codependent drug-companion diagnostics technologies in improving patient outcomes [23]. EPEMED's short-term goals include identifying ways to facilitate development of personalized treatments and ensure they are made available to patients quickly and cost-effectively, establish clear regulatory guidance on development of diagnostic test and personalized drug therapy, and improve Europe-wide market access for high value companion diagnostics [24]. In 2013, EPEMED began collaborations with numerous other European organizations including EuroBioForum, EuroMeDiag, Eurobiomed, La Charite, and more to launch major market studies on personalized medicine patient's access in Europe. The group has held annual international conferences since its inception and has released multiple studies including the recent EPEMED–The Office of Health Economics (OHE) study "The value of knowing and knowing the value: improving health technology assessment of complementary diagnostics" in 2016 [25].

cross-department partnership will support a centralized commissioning that will provide new incentives such as increasing budgetary capabilities, finding new funding, and providing training and education [22]. "The recent Accelerated Access Review proposals, with their strong emphasis on precision medicine, address several long-recognized challenges in the UK translational and commissioning landscape, and have the potential to solidify the UK's leadership," says Iain Miller, Ph.D., Founder, Healthcare Strategies Group, a UK-based life sciences consultancy.

As new targeted therapies are developed, the United Kingdom must ensure that patients have access to these lifesaving therapies as well as the diagnostic tests that must be conducted before patients can qualify for these therapies. In the United Kingdom, a positive recommendation from NICE implies mandatory funding for the NICE-recommended drug as well as its associated companion diagnostics, but local budget holders are not provided with additional funding. As a result, budget constraints can limit or delay access to drugs, even if they are recommended by NICE. To ensure that patients benefit from the many precision medicine advances in the United Kingdom, mandatory funding for drugs that are recommended by NICE must be extended to their associated companion diagnostics as well; this is the next challenge to be overcome [26].

Indeed, in every country across the globe, not just in the United Kingdom, facilitating reimbursement and access to both targeted therapies and their companion diagnostics will remain a key challenge and has the potential to grow increasingly problematic as more and more expensive therapies and companion diagnostics become available. Fortunately, costs can fall dramatically over time as seen with the dramatic decrease in the cost of genome sequencing from $3,000,000 per whole genome in 2003 to less than $1,500 today. This highlights the need to not only focus on using technology to develop brand new therapies but also develop and use technology to bring down costs so that medical advances are more accessible to all.

Germany

Germany, which is Europe's largest pharmaceutical market [26], also played a significant role in the Human Genome Project. Now that the cost and time associated with genetic sequencing has plummeted, obtaining genetic data is no longer the primary challenge, and Germany has shifted its focus to the interdisciplinary evaluation of patient data. Germany's Centre for Personalised Medicine, known in German as the Zentrum für Personalisierte Medizin (ZPM), was established in January 2015 to bring together 23 departments and institutes of the University of Tübingen's medical faculty to improve disease diagnosis and develop personalized treatments for patients in a variety of disease areas. Although Germany has a handful of other centers focused on precision medicine, most of these centers focus on only one specific topic within precision medicine, whereas the ZPM brings together many fields of study, including genetics, imaging, computer science, and engineering. This allows the ZPM to consider genetic material (genomics), proteins (proteomics), metabolic processes (metabolomics), and even imaging data when making diagnoses and treatment decisions [27].

According to Dr Nisar Malek, director of the ZPM, "The common belief that personalised medicine will lead to dramatically rising costs is definitely not right. Just think about the horrendous cost of medicines and treatments that have little or no effect; therapies tailored to the requirements of individual patients would avoid such costs… A cancer such as pancreatic cancer involves a large number of different cells, and we basically need to treat many diseases simultaneously on the molecular level. Therefore, knowledge and experience from widely differing research fields need to be integrated in order to develop new, individual treatments" [26]. If the ZPM and similar initiatives are successful, precision medicine may have the potential to not only improve health outcomes but also decrease overall healthcare costs.

Many German labs have also identified another way to decrease costs. Pathologists in German labs often note that developing their own tests, which are called lab developed tests (LDTs), can be a more cost-effective way to conduct molecular testing even when companion diagnostic testing kits are already available. Although labs require time and money to develop LDTs, a lab with sufficiently high testing volume can ultimately save money by using LDTs rather than buying kits from diagnostic manufacturers [10]. Labs in Germany that do not use approved kits can be reimbursed for conducting tests because codes for molecular diagnostics in Germany typically refer to the testing method used (e.g., amplification or sequencing) or to the biomarker that is tested [28].

While LDTs have worked well in many German labs, not all countries embrace the use of LDTs. In the United States, LDTs have been met with significant regulatory and reimbursement questions, covered in more detail in the chapters ahead. Using approved companion diagnostic kits helps to ensure that all patient samples are tested in a standardized way. If the kit is approved (e.g., CE marked in Europe or FDA approved in the United States), then patients' test results from the kits are expected to be relatively reliable. However, using an LDT decreases the amount of standardization and can result in false negatives (i.e., patients who are biomarker positive may not be identified) or false positives (i.e., patients who are biomarker negative may be incorrectly identified as biomarker positive and consequently receive a targeted therapy from which they are likely to see little benefit). While approved companion diagnostic kits can also result in some false positives and false negatives, limited oversight over LDTs can result in a much greater proportion of inaccurate results [29].

Germany continues to innovate even at the individual lab level and brings new meaning to German engineering. Germany's innovation is not limited to its diagnostic labs or universities. Qiagen, a global leader in companion diagnostic development and commercialization with annual sales exceeding $1.25 billion, was founded in Dusseldorf, Germany, in 1984 [30]. Peer Schatz, Chief Executive Officer of Qiagen, believes that a key to their success is innovation, which in turn is driven by selection, development, and encouragement of its employees. As Mr. Schatz commented after Qiagen won one of several innovation awards they have received over the years, "We are very proud to have won this year's 'Best Innovator' Award as it recognizes the efforts we undertake to lead the field through innovation. We are always working to further expand and reinforce our market and technology leadership. In the biotechnology industry, the key to success can only be innovation. Our main success factor is that we place selection, development and encouragement of employees at the center of our innovation management strategy" [31]. Germany's commitment to innovation and its ambition to improve healthcare while lowering costs provide valuable lessons as other countries develop their precision medicine capabilities.

Europe, particularly France, the United Kingdom, and Germany, excels at both technological innovation in the field of precision medicine and developing mechanisms to promote access to these innovations. In these countries, we find innovation centers such as Genopole, Genomics England, and the ZPM. In addition, each of these countries is working to overcome the challenges of access through the INCa platform, universal healthcare, and LDTs. Each of these countries, and indeed others across Europe including Sweden and Iceland, showcases the progress that has already been made in precision medicine, and each has established goals that show us how much more can be achieved.

References

1 Adapted from: Nowak F, Calvo F, Soria J. Europe does it better: molecular testing across a national health care system – the French example [Internet]. ASCO University; 2013 [cited Jan 3, 2017]. Available from: http://meetinglibrary. asco.org/content/121-132

2 United Nations, Department of Economic and Social Affairs. World population prospects: the 2015 revision, key findings and advance tables [Internet]. United Nations; 2015 [cited Jan 3, 2017]. Available from: https://esa.un.org/unpd/wpp/publications/files/key_findings_wpp_2015.pdf

3 World economic outlook database [Internet]. International Monetary Fund; [cited Jan 3, 2017]. Available from: http://www.oecd-ilibrary.org/content/data/data-00540-en

4 Ferlay J, Steliarova-Foucher E, Lortet-Tieulent J, Rosso S, Coebergh JWW, Comber H, et al. Cancer incidence and mortality patterns in Europe: estimates for 40 countries in 2012. Eur J Cancer. 2013;49(6):1374–403.

5 Eurostat. Cancer related healthcare personnel and equipment, 2008 and 2013 (per 100,000 inhabitants) [Internet]; 2015 [cited Apr 6, 2017]. Available from: http://ec.europa.eu/eurostat/statistics-explained/index.php/Cancer_statistics

6 Goss P. The Lancet Oncology: commission shows good progress in cancer care in Latin America. EurekAlert [Internet]. AAAS; Oct 28, 2015 [cited Apr 6, 2017]. Available from: https://www.eurekalert.org/pub_releases/2015-10/tl-tlo102715.php

7 Garfield D, Brenner H, Lu L. Practicing Western Oncology in Shanghai, China: one group's experience. J Oncol Pract. 2013 Jul; 9(4):e141–e144.

8 Chevreul K, Brigham KB, Durand-Zaleski I, Hernández-Quevedo C. Health systems in transition: France health system review [Internet]. The European Observatory on Health Systems and Policies; 2015 [cited Jan 3, 2017]. Available from: http://www.euro.who.int/__data/assets/pdf_file/0011/297938/France-HiT.pdf

9 Le programme de dépistage organisé [Internet]. Institut National Du Cancer; Sep 25, 2015 [cited Dec 7, 2016]. Available from: http://www.e-cancer.fr/

Professionnels-de-sante/Depistage-et-detection-precoce/Depistage-du-cancer-du-sein/Le-programme-de-depistage-organise

10 OECD. Health Status [Internet]. Paris: Organisation for Economic Co-operation and Development; Oct 2016 [cited Dec 7, 2016]. Available from: http://www.oecd-ilibrary.org/content/data/data-00540-en

11 Gauvreau D. Some thoughts on clusters staying connected to R&D expertise and feeding innovation but no solutions [Internet]. Genopole; Dec 10, 2015 [cited Dec 7, 2016]. Available from: http://www.lifescience-cluster-innovations.space/wp-content/uploads/2015/10/Global-Cluster-Hub-Genesis-Genopole.pdf

12 Catapult Precision Medicine. Vision and mission [Internet]. Catapult Precision Medicine; [cited Jan 4, 2017]. Available from: https://pm.catapult.org.uk/about-us/vision-and-mission/

13 Genomics England. The 100,000 genomes project [Internet]. Department of Health; [cited Jan 5, 2017]. Available from: https://www.genomicsengland.co.uk/the-100000-genomes-project/

14 Genomics England. First patients diagnosed through the 100,000 genomes project [Internet]. Department of Health; Mar 11, 2015 [cited Jan 5, 2017]. Available from: https://www.genomicsengland.co.uk/first-patients-diagnosed-through-the-100000-genomes-project/

15 Mapping the UK precision medicine landscape [Internet]. Knowledge Transfer Network; [cited Jan 4, 2017]. Available from: http://pmlandscape.ktn-uk.org/

16 Amos J. Plan to "Catapult" UK space tech [Internet]. BBC News; Jan 4, 2012 [cited Jan 12, 2017]. Available from: http://www.bbc.com/news/science-environment-16409746

17 The Catapult Programme [Internet]. Catapult; [cited Jan 12, 2017]. Available from: https://catapult.org.uk/

18 Diagnostic Evidence Co-operative London [Internet]. NHS Diagnostic Evidence Co-operative London; [cited Jan 12, 2017]. Available from: http://london.dec.nihr.ac.uk/

19 About us [Internet]. Diagnostic Evidence Co-operative London; [cited Jan 12, 2017]. Available from: http://london.dec.nihr.ac.uk/about-us/

20 Consultation on changes to technology appraisals and highly specialised technologies [Internet]. National Institute for Health and Care Excellence; [cited Jan 12, 2017]. Available from: https://www.nice.org.uk/about/what-we-do/our-programmes/nice-guidance/nice-technology-appraisal-guidance/consultation-on-changes-to-technology-appraisals-and-highly-specialised-technologies

21 Patients to get faster access to the most cost effective treatments under proposed changes to NICE process [Internet]. NHS England; [cited Jan 12, 2017]. Available from: https://www.england.nhs.uk/2016/10/proposed-changes/

22 Accelerated access review: final report [Internet]. Wellcome Trust; Oct 24, 2016 [cited Jan 12, 2017]. Available from: https://www.gov.uk/government/publications/accelerated-access-review-final-report

23 Huriez A. Make personalised medicine and diagnostics a reality for European patients [Internet]. EPEMED; [cited Jan 12, 2017]. Available from: http://www.epemed.org/online/www/content/79/80/ENG/index.html

24 Near term goals [Internet]. European Personalised Medicine Diagnostics Association (EPEMED); [cited Jan 12, 2017]. Available from: http://www.epemed.org/online/www/content/79/81/ENG/index.html#ribbon

25 Release of EPEMED OHE Study 2016 [Internet]. European Personalised Medicine Diagnostics Association (EPEMED) and Office of Health Economics; Jul 5, 2016 [cited Jan 12, 2017]. Available from: http://www.epemed.org/online/www/content2/104/105/ENG/5115.html

26 Pharmaceutical industry [Internet]. Germany Trade & Invest; [cited Jan 5, 2017]. Available from: http://www.gtai.de/GTAI/Navigation/EN/Invest/Industries/Life-sciences/pharmaceuticals.html

27 Centre for Personalised Medicine in Tübingen—developing tailor-made treatments for patients [Internet]. BIOPRO Baden-Württemberg; Jun 15, 2015 [cited Jan 5, 2017]. Available from: https://www.gesundheitsindustrie-bw.de/en/article/news/centre-for-personalised-medicine-in-tuebingen-developing-tailor-made-treatments-for-patients/

28 Bücheler M, Brüggenjürgen B, Willich S. Personalized medicine in Europe – enhancing patient access to pharmaceutical drug-diagnostic companion products [Internet]. European Personalised Medicine Diagnostics Association (EPEMED); Nov 2014 [cited Jan 5, 2017]. Available from: http://www.epemed.org/online/www/content2/104/107/910/pagecontent2/4339/791/ENG/EpemedWhitePaperNOV14.pdf

29 Ray T. FDA holding off on finalizing regulatory guidance for lab-developed tests [Internet]. GenomeWeb; Nov 2016 [cited Jan 5, 2017]. Available from: https://www.genomeweb.com/molecular-diagnostics/fda-holding-finalizing-regulatory-guidance-lab-developed-tests

30 History [Internet]. QIAGEN; [cited Jan 6, 2017]. Available from: https://www.qiagen.com/be/about-us/who-we-are/history/

31 QIAGEN. QIAGEN selected as the company with the best innovation management strategy in Germany [Internet]. PRNewswire; Jun 9, 2011 [cited Jan 6, 2017]. Available from: http://www.prnewswire.com/news-releases/qiagen-selected-as-the-company-with-the-best-innovation-management-strategy-in-germany-123553644.html

Part 2

The Present

4

Our Reality Today

The Patient Journey in Precision Medicine

Some of the most incremental advancements in precision medicine to date have taken place within the cancer industry. These oncology developments have been especially felt in the fields of solid tumors, or growing tumors within the skin, organs, or bone. This differs from blood tumors, or hematological malignancies, and as such there are different and important nuances to understand when considering precision medicine for solid tumors.

Getting a bit technical, a tumor is a cancerous mass of cells within the body. This mass of cells is different from normal cells within the body, as it is growing at an incredibly rapid pace [1]. This growth is fueled by chemical reactions and rapid metabolism of fats and carbohydrate stores, which is one reason why an early indicator of cancer development is weight loss in patients. When the tumor is just starting to grow, it will typically develop in one location. This could be within the prostate, the liver, the skin, the brain, or basically anywhere within the body. When the tumor is located in just one isolated location, it can be considered "localized," or stage 1. As the tumor continues to grow and expand, tumor cells can spread beyond the initial site and manifest elsewhere within the body. Depending upon how dramatic the expansion of the tumor happens to be, the cancer is then "staged" (as part of the "diagnosis" portion of the diagnostics continuum explained in a previous chapter), with stage 4 cancer indicating the most severe widespread additional tumors throughout the body [2]. These additional tumors, also known as growths or lesions, are considered additional metastases (growths away from the tumor of origin), which is why the late-stage condition is known as "metastatic" cancer. There are many important elements to the progression of cancer, but for now all we need to know is that stage 1 is early and local and stage 4 is late and metastatic.

To better introduce and understand the oncology industry, it is important to outline and describe the experience that a patient has throughout the course of their disease. Oftentimes, this is referred to as the "patient journey." I started this process in the first two chapters by outlining the diagnostics portion of the journey (the diagnostics continuum), but there is much more to the patient

Personalizing Precision Medicine: A Global Voyage from Vision to Reality, First Edition.
Kristin Ciriello Pothier.
© 2017 John Wiley & Sons, Inc. Published 2017 by John Wiley & Sons, Inc.

journey than the diagnosis. By learning more about the patient journey, we're able to learn more about the medical steps that the patient goes through from the initial detection and diagnosis of the disease through the treatment, monitoring, and potential remission of the cancer.

Since cancer is an unfortunately broad disease, with potential cancerous growth occurring just about anywhere throughout the body, I will focus on two specific cases to develop an understanding of the patient journey. In order to best illustrate the journey, as well as the way precision medicine has influenced its progression over the last several decades, I will start with one of the most common oncology areas: breast cancer.

Breast Cancer

In 2000, after several developments in the field and after many women had lost their lives to breast cancer, the FDA and other industry regulators instituted guidance for early detection through an emerging medical procedure called digital mammograms [3]. These digital mammograms at the time were novel systems for screening and diagnosing breast cancer, beyond what was possible with physical inspection to detect growths, or lumps. The mammography procedure also includes an X-ray component, a type of advanced imaging diagnostics, as the compressed breasts are analyzed for growths. While the recommended utilization has evolved over time, the most recent guidance put forward by the US Preventive Services Task Force (USPSTF) recommends that women with an average risk of breast cancer only need testing every 2 years starting at age 50 [4]. The process to obtain a mammogram is fairly common, as it is completed typically in a basic imaging center and takes less than 15 minutes to perform the examination [5]. The results are interpreted by a radiologist, a specialist who analyzes X-rays and other imaging tests, shortly after the diagnostic procedure. Results are typically passed along within a few weeks.

For women with positive mammograms, this next imaging procedure is a more detailed diagnostics mammogram, an ultrasound, to better assess the lump and understand whether there are additional growths in the breast tissue or in the surrounding tissue and organs [6]. If the diagnostic mammogram results are positive or suspicious, the next step is for the woman to have additional imaging tests done, such as an MRI. This procedure is also completed by a radiology technician and can take place in a hospital or outpatient clinic or imaging center [7].

Then, if the results from the more comprehensive diagnostic mammogram are positive and still suspicious, it is necessary to understand whether the growths within the breast tissue are actually cancerous (malignant) or if they are not immediately threatening (benign) [8]. The process to remove tissue from the breast is called a biopsy and can be taken either with a needle (called a core biopsy) or with a traditional blade or scalpel (a surgical biopsy). Depending on the location of the suspicious growth, the physician will recommend one or the other. If a surgical biopsy is required, then a surgical oncologist is necessary. Otherwise, medical oncologists are equipped to perform the core needle biopsy [9]. The biopsy is taken using advanced imaging technologies, often guided with an ultrasound or an MRI, to make sure that it is taken directly from the abnormal tissue in order to reduce the amount of excess tissue removed from the breast.

Laboratory diagnostic tests are then required to assess whether the growth is cancerous. In order to run these tests, after the biopsy is taken, the physician needs to make sure that the sample is preserved for testing. To do this, as soon as the biopsy is taken from the patient, a surgeon/physician will send the sample to the laboratory to preserve it for as long as possible. Once the sample gets to the laboratory, a physician specially trained in the preservation and analysis of tissue, called a pathologist, takes a portion of the sample and examines it under a microscope. This process is known as anatomical pathology, or the examination of a tumor to determine whether it is cancerous or benign [10]. Before or after this initial examination, the pathologist will preserve the sample by covering it in a waxy substance called paraffin, which prevents the tissue from drying up [11]. This allows the pathologist to run additional tests later if needed.

If the pathologist determines that the tissue is cancerous after examining it under the microscope, then this is when the patient officially is diagnosed with cancer. Up until this point, there is a chance that the mass of cells is benign, or not harmful at all. This is why for many women, the process of getting a biopsy is often misconstrued as a diagnosis of cancer, but until the cells are looked at under a microscope, nothing is certain.

Now, if the pathologist determines that the cells are in fact cancerous, then there are three tests that are commonly performed on the sample [12]. These additional tests, or downstream analyses following the initial physical inspection, are the reasons why the pathologist will only use a small portion of the sample for the initial pathology testing. The first two tests that are performed on the sample are for increased activity of two different types of hormone receptors, estrogen receptor (or ER) and progesterone receptor (or PR) [13]. The pathologist will use specialty tests that involve staining the tissue and examining it under a microscope again to perform this analysis. This method is called immunohistochemistry. Breaking this word down, immuno- refers to the antibodies that bind to specific antigens to create colors used in the staining process, and histo- means tissue [14].

The third test that a pathologist will perform is a HER2/neu test. This test looks for a specific protein, or a gene that is involved in the development of that specific protein. The gene assessed is formally called the human epidermal growth factor receptor 2, and it is responsible for making the HER2 protein [15].

Based on the results of these three tests, the pathologist's report back to the medical oncologist can have several variations, depending on the positive or negative expression of the three tests (Figure 4.1). Thus, there are eight discrete possibilities based on the genetic tests.

These tests are important because they will help the medical oncologist determine which course of action to take with the patient. But the genetic test is not the only form of testing that is considered when making a treatment recommendation. There are also the results of the imaging tests to indicate where the (now determined to be cancerous) abnormal tissue exists. This analysis of the degree of expansiveness of the cancer takes place across three other variables—the size of the tumor (T), whether the tumor has spread to additional lymph glands or nodes (N), and whether the cancer has metastasized (M) (Figure 4.2) [16]. The combination of these three variables determines the severity, or staging, of the cancer.

As you can see, getting diagnosed with "breast cancer" doesn't really tell a patient much at all. In order for the physician to make a decision regarding the best course of action for the patient, each of these factors needs to be carefully weighed in order to make the appropriate diagnosis. Since we have four striations for tumor size, three different gene markers, two options for node involvement, and two options for degree of metastasis, this results in nearly 200 possible differential diagnoses of "breast cancer"! In addition, if there is potential family history of breast cancer (and sometimes even when there isn't), a physician may order a different type of test, a blood or saliva test, to specifically determine whether the woman carries a BRCA mutation in one of two BRCA tumor suppressor genes in order to further assess how severe the woman's cancer is or could be; BRCA mutations carry significantly increased risk of breast and ovarian cancer. In 2013, actress Angelina Jolie made these

ER	PR	HER2	Probability (%)
+	+	+	75–80
+	+	−	
+	−	+	40–50
+	−	−	
−	+	+	25–30
−	+	−	
−	−	+	<10
−	−	−	

Figure 4.1 The different possible outcomes of ER, PR, and HER2 tests.

Stage	Tumor size (T)	Lymph node involvement (N)	Metastasis (M)
Stage I	<2 cm	No	No
Stage II	2–5 cm	No, or same side of breast	No
Stage III	>5 cm	Yes, same side of breast	No
Stage IV	N/A	N/A	Yes

Figure 4.2 Definitions of different stages of breast cancer.

mutations and the testing for them famous when she chose a prophylactic mastectomy after she learned the results of her BRCA test. She had an 87% risk of breast cancer and a 50% risk of ovarian cancer, thus leading her to choose a prophylactic mastectomy and, 2 years later, removal of her ovaries and fallopian tubes as well [17, 18].

Once the different physicians have appropriately diagnosed the type of breast cancer a patient has, the next step is to consider treatment options. The course of action for treatment will almost always include one or more elements of surgical (cutting) and medical (drug) treatment (Figure 4.3) [19].

We've now identified two different types of surgical treatments and four different types of medical treatments, all of which can be used independently or in addition to the others. This means that there are at least 64 different potential treatment strategies for treating breast cancer, and that's not even accounting for the type of drug or the amount of drug that is used.

In order to help sort the options, the American Society of Clinical Oncology (ASCO) and the National Comprehensive Cancer Network (NCCN), together with most major medical centers, have developed pathways pointing to the treatments with the most success in their populations. Most physicians will consult one if not two different sets of pathways to guide her patient's own path, again with the help of diagnostics to make the decisions more clear. A now standard test in the United States called Oncotype DX, offered by the company Genomic Health, uses the tumor's signature of 21 different genes to determine, after surgery, what a woman's risk of recurrence is and whether adding chemotherapy will decrease that risk. Women are given a recurrence score and classified into low (l<18), intermediate (18–30), or high risk (31 or more). Low risk shuttles women into the option of no chemotherapy, while high risk shuttles them into chemotherapy. The intermediate zone, one I have myself described as "the infuriatingly confusing zone" after one of my best friends was stuck squarely in that category, is one where the benefit of chemotherapy needs discussion with one's physician. Other factors, such as patient age, patient concern and preference, comorbidities, conditions that may increase the risk of chemotherapy-associated toxicity, tumor size and grade, and degree of ER expression, are considered at that point [20]. My friend chose the aggressive chemotherapy

Option	Method	Description
Surgical	Lumpectomy	Surgical removal of the cancerous cells
	Mastectomy	Surgical removal of entire breasts
Medical	Radiation therapy	Use of high-energy particles to destroy cancer cells
	Chemotherapy	Infusion to kill cancer cells and prevent growth
	Hormonal therapy	Shrinks and prevents recurrence of ER+ or PR+ tumors
	Targeted therapy (precision medicine)	Slows and stops growth of HER2+ tumors or other gene-based mutation tumors; patient must have specific mutated tumors in order to respond

Figure 4.3 Six different treatment options for breast cancer patients.

route after a double mastectomy because of her young age and overall good health (other than the invasive breast cancer she was diagnosed with!). Still more complex is the newest, most comprehensive way to diagnose patients and their potential treatment journey, and that is through comprehensive genomic profiling (CGP). CGP, such as FoundationOne® offered by Roche, analyzes the entire coding sequence of 315 cancer-related genes plus select introns from 28 genes often rearranged or altered in solid tumor cancers [21], allowing for the most complete information on all the possible mutations a patient's tumor may have, and how they may respond to a number of targeted therapies. In some patients, such as those with triple negative breast cancer (ER negative, PR negative, and Her2 negative), additional sequencing can discover other information to best guide therapy. In a percentage of triple negative patients, they actually have a base mutation that changes the status of their Her2– to Her2+, resulting in the ability to be treated by targeted therapy Herceptin when prior to the additional sequencing they had no more options [22].

Once the diagnosis continuum is complete, the patient will cycle through her treatment journey. As noted previously, there are multiple combinations of therapy depending on the tumor(s), the stage, and the guidelines the physician follows. Additionally, the patient may be eligible for clinical trials with drugs that are still in the testing phase, which makes the treatment paradigm more complex, and then there are possibilities of emerging therapies—like immunotherapy—that are only used as a last resort in breast cancer today. The following figure gives a simplified version of what the physician and the patient face when making decisions on the course of therapy for breast cancer (Figure 4.4).

Breast cancer is one of the most commonly diagnosed and treated forms of cancer, which is why several generations of research of medications have been developed to treat patients at various phases of the disease. However, not all cancers are like this. Unfortunately, the options for potential therapies get slimmer when patients are diagnosed later in the disease. One type of cancer that's often not diagnosed early is late-stage skin cancer, also known as metastatic melanoma. Although patients diagnosed with metastatic melanoma are few, just 2% of all melanoma patients, that number is increasing steadily over time, creating a need for better therapies [24].

Metastatic Melanoma

One of the first interviews I did for this book was with a dear friend of my parents, whose husband had died of metastatic melanoma a few years before. Tommy was one of my father's best friends from "the old neighborhood" and

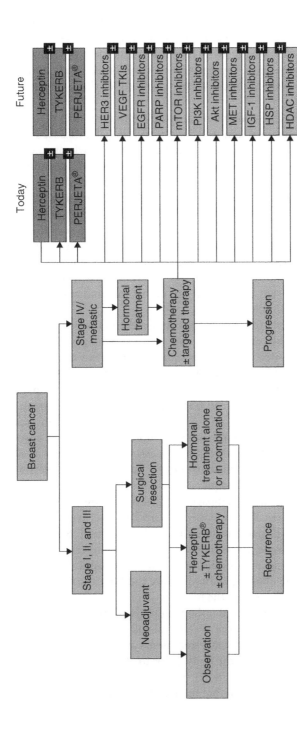

Figure 4.4 Simplified breast cancer patient journey today and in the future [23].

was one of those larger-than-life people who always seemed to do whatever he wanted and find a way to make it successful for him. His wife Molly was a fabulous sidekick, moving with him and their kids all around the United States every few years to try one thing or the other. A melanoma diagnosis was surprising to them, but it wasn't going to let them down. And so their journey began.

The path to diagnosis of metastatic melanoma can take place in two ways. Either a patient can be diagnosed earlier, and then the disease can progress to the metastatic (stage IV) phase, or a patient can be newly diagnosed as a stage IV patient (like Tommy). Newly diagnosed metastatic melanoma is a very dangerous but very real form of the disease. The diagnosis of metastatic melanoma begins with a self-observed strangely-shaped mole or growth on the skin. While many people disregard these growths, when they either grow, change shape, or cause pain and irritation, people often seek the second opinion of a physician [25]. The physician most commonly consulted is a dermatologist, who in addition to treating cosmetic skin issues is the first line of defense for metastatic skin cancer. The dermatologist will perform a physical exam to note the size, shape, and texture of the area and will check nearby lymph nodes for possible growths.

Once the physical exam has been completed and the dermatologist suspects the growth might be cancerous, he will take a skin biopsy. This process is usually not very complicated, as taking a biopsy from the skin is much easier to access compared with, say, a lung tumor or a kidney tumor. For skin biopsies, a local pain-numbing agent, or anesthetic, is used prior to the biopsy. Depending on the physician, the biopsy can be taken by using a punch biopsy, with a specific round cutting tool, or an incisional biopsy, where a physician will cut deeper into the skin to understand if the mole has penetrated beyond the surface. From here, the biopsy will be processed the same way as other tumor types, with physical examinations under a microscope to determine whether the cells are cancerous or benign.

For many with newly diagnosed metastatic melanoma, the dermatologist performing the physical examination will also uncover lymph node growths, or metastases, and will perform biopsies of those as well. Similar to the earlier breast cancer example, these biopsies are a bit more tricky and invasive and require more precision. Additionally, once metastatic melanoma has been detected, the same imaging techniques are used to understand just how much the tumor has spread within the body.

So for metastatic melanoma, the staging is simple. The cancer has spread beyond a single site, and surgical removal will not help the patient. While this is unfortunate, there are medical treatments available to stop or slow the proliferation of tumors within the body. Prior to 2011, only two medications were approved to treat the metastatic form of skin cancer: dacarbazine (approved in 1975), a chemotherapy agent that has just a 13% chance of shrinking the tumor, and high-dose IL-2 (approved in 1998), a high-dose immune-boosting therapy that was found to be curative in about 4% of patients but has very serious and possibly fatal side effects [26]. These drugs, as limited and ineffective as they may seem, were the only two options for metastatic melanoma patients for over 10 years. Molly took out her old notes for me to go over Tommy's journey, which occurred in the late 1990s and early 2000s. She honestly and at times tearfully told me how they tried the typical treatments and then searched for clinical trials and how Tommy worked (at this point, as a teacher) up until his legs swelled so much he could no longer get to class. But in the end, there was nothing left for him.

Fast forward to 2011, when the entire metastatic melanoma treatment guidelines were turned upside down with the approval of the first targeted therapy for metastatic melanoma patients with BRAF mutations (BRAF+) on the V600E gene. A pivotal phase III trial of this drug, later named Zelboraf, showed that a positive response rate could be achieved in 48% of patients compared with 5% for dacarbazine (approved in 1975) [27]. This approval brought hope to thousands of metastatic melanoma patients and brought targeted therapies to the forefront of metastatic melanoma treatment. In 2016, we now have several more options available for metastatic melanoma, which can be used individually or in combination (Figure 4.5) [28].

So let's think about this. From 1975 to 2011 we had just about no improvements for these metastatic melanoma patients. But then in a span of just 5 years, we saw seven novel and life-changing therapies come to market. How did this happen? It was all due to precision medicine. It was all due to finally understanding the genetic causes of the disease and developing drugs and medications to specifically target these issues, as opposed to using widespread chemotherapy to kill cancer cells (while also killing important living cells at the same time). And we're just now scratching the surface of precision medicine, with hundreds of millions of dollars being invested each year to better understand the genes that cause tumor growth and to determine which patients should receive which drug combinations. As I mentioned in the Introduction, in 2015, 28% of all drugs approved by the FDA had biomarker information in

Drug	Approval year	Type
Dacarbazine	1975	Chemotherapy, not targeted
Proleukin	1998	High-dose IL-2
Zelboraf (vemurafenib)	2011	BRAF inhibitor
Yervoy (ipilimumab)	2011	Anti-CTLA-4 Immune checkpoint inhibitor
Tafinlar (dabrafenib)	2013	BRAF inhibitor
Mekinist (trametinib)	2014	MEK inhibitor
Keytruda (pembrolizumab)	2014	Programmed death 1 (PD-1) protein inhibitor
Opdivo (nivolumab)	2014	Programmed death 1 (PD-1) protein inhibitor
Cotellic (cobimetinib)	2015	MEK inhibitor

Figure 4.5 The various drugs available to treat metastatic melanoma.

their labels [29]. Analysis of drugs in development suggests that this percentage will rise significantly in the coming years, and each and every one of those targeted drugs has a diagnostic that can open the door to the right drug. Unfortunately, this is too late for Tommy, or my grandmother who hid a "freckle" on her arm for years before being diagnosed late stage and having nothing to help her, or all of the other patients before this age of precision medicine who ended their journeys too soon. At its core, no matter where in the world you are, precision medicine is better understanding of the patient throughout the course of his disease. Understanding those patients, from detection to treatment to monitoring, across the ENTIRE patient diagnostic, therapeutic, and service journey, is the central theme to personalizing precision medicine.

References

1 What are tumors? [Internet]. The Sol Goldman Pancreatic Cancer Research Center at Johns Hopkins Medicine; [cited Dec 6, 2016]. Available from: http://pathology.jhu.edu/pc/BasicTypes1.php?area=ba

2 Staging [Internet]. National Cancer Institute; Mar 9, 2015 [cited Dec 6, 2016]. Available from: https://www.cancer.gov/about-cancer/diagnosis-staging/staging

3 FDA permits marketing of first direct-to-consumer genetic carrier test for Bloom syndrome [Internet]. U.S. Food & Drug Administration; Feb 19, 2015 [cited Nov 22, 2016]. Available from: http://www.fda.gov/NewsEvents/Newsroom/PressAnnouncements/ucm435003.htm

4 Final update summary: breast cancer: screening [Internet]. US Preventive Services Task Force; Sep 2016 [cited Dec 6, 2016]. Available from: https://www.uspreventiveservicestaskforce.org/Page/Document/UpdateSummaryFinal/breast-cancer-screening1

5 Screening mammogram [Internet]. Memorial Hermann; [cited Nov 22, 2016]. Available from: http://www.memorialhermann.org/imaging-and-diagnostics/screening-mammogram/

6 Mammogram [Internet]. National Breast Cancer Foundation; [cited Dec 6, 2016]. Available from: http://www.nationalbreastcancer.org/diagnostic-mammogram

7 Breast MRI [Internet]. Cancer.Net; Sep 2016 [cited Dec 6, 2016]. Available from: http://www.cancer.net/navigating-cancer-care/diagnosing-cancer/tests-and-procedures/breast-mri

8 Breast biopsy [Internet]. American Cancer Society; [cited Dec 6, 2016]. Available from: http://www.cancer.org/treatment/understandingyourdiagnosis/examsandtestdescriptions/forwomenfacingabreastbiopsy/

9 Bennett M. Breast core biopsy [Internet]. InsideRadiology; Nov 11, 2016 [cited Dec 6, 2016]. Available from: http://www.insideradiology.com.au/breast-core-biopsy/

10 Anatomic pathology [Internet]. Lab Tests Online; Jul 22, 2014 [cited Dec 6, 2016]. Available from: https://labtestsonline.org/understanding/features/anatomic-pathology/

11 Preserving breast tissue samples for pathology [Internet]. Susan G. Komen®; Oct 19, 2016 [cited Dec 6, 2016]. Available from: http://ww5.komen.org/BreastCancer/Preserving-Breast-Tissue-Samples-for-Pathology.html

12 IHC tests (ImmunoHistoChemistry) [Internet]. Breastcancer.org; Oct 23, 2015 [cited Dec 6, 2016]. Available from: http://www.breastcancer.org/symptoms/testing/types/ihc

13 Estrogen and progesterone receptor testing for breast cancer [Internet]. Cancer.Net; Apr 19, 2010 [cited Dec 5, 2016]. Available from: http://www.cancer.net/research-and-advocacy/asco-care-and-treatment-recommendations-patients/estrogen-and-progesterone-receptor-testing-breast-cancer

14 Ramos-Vara JA, Miller MA. When tissue antigens and antibodies get along revisiting the technical aspects of immunohistochemistry—the red, brown, and blue technique. Vet Pathol. Jan 2014;51(1):42–87.

15 Moynihan TJ. HER2-positive breast cancer: what is it? [Internet]. Mayo Clinic; Mar 25, 2015 [cited Dec 6, 2016]. Available from: http://www.mayoclinic.org/breast-cancer/expert-answers/faq-20058066

16 The American Cancer Society medical and editorial content team. How is breast cancer staged? [Internet]. American Cancer Society; Jun 1, 2016 [cited Dec 6, 2016]. Available from: http://www.cancer.org/cancer/breastcancer/detailedguide/breast-cancer-staging

17 Jolie A. My medical choice [Internet]. The New York Times; May 14, 2013 [cited Dec 6, 2016]. Available from: http://www.nytimes.com/2013/05/14/opinion/my-medical-choice.html

18 Angelina Jolie Pitt's story highlights BRCA testing and personal health decisions [Internet]. LabTestsOnline. Apr 9, 2015 [cited Dec 6, 2016]. Available from: https://labtestsonline.org/news/150409brca/

19 Cancer.net Editorial Board. Breast cancer: treatment options [Internet]. Cancer.Net; Jun 25, 2012 [cited Dec 6, 2016]. Available from: http://www.cancer.net/cancer-types/breast-cancer/treatment-options

20 The Recurrence Score report [Internet]. Intermediate Oncotype DX; [cited Dec 6, 2016]. Available from: http://intermediate.oncotypedx.com/en-US/The-Recurrence-Score-Result

21 FoundationOne® website [Internet]. FoundationOne®; [cited Dec 6, 2016]. Available from: http://www.comprehensivegenomicprofiling.com/#increase-role

22 Palma NA, Chalmers Z, Li Y, Bailey M, Ross JS, Balasubramanian S. A U.S.-based prospective, multi-center, non-interventional study of the role of comprehensive genomic profiling in the clinic. J Clin Oncol 2015;33(15_suppl):e22183.

23 Adapted from: Pothier K, Gustavsen G. Combatting complexity: partnerships in personalized medicine. Pers Med. 2013;10(4):387–96.

24 Cancer Research UK. Skin cancer incidence statistics [Internet]. Cancer Research UK; May 15, 2015 [cited Dec 60, 2016]. Available from: http://www.cancerresearchuk.org/health-professional/cancer-statistics/statistics-by-cancer-type/skin-cancer/incidence

25 The American Cancer Society Medical and Editorial Content Team. Tests for melanoma skin cancer [Internet]. The American Cancer Society; May 19, 2016 [cited Dec 5, 2016]. Available from: http://www.cancer.org/cancer/skincancer-melanoma/detailedguide/melanoma-skin-cancer-diagnosed

26 Smyth EC, Carvajal RD. Treatment of metastatic melanoma: a new world opens [Internet]. Skin Cancer Foundation; [cited Dec 5, 2016]. Available from: http://www.skincancer.org/skin-cancer-information/melanoma/melanoma-treatments/treatment-of-metastatic-melanoma

27 Rebecca VW, Sondak VK, Smalley KS. A brief history of melanoma: from mummies to mutations. Melanoma Res. Apr 2012;22(2):114–22.

28 FDA approved drugs for melanoma [Internet]. AIM at Melanoma Foundation; [cited Dec 6, 2016]. Available from: https://www.aimatmelanoma.org/melanoma-treatment-options/fda-approved-drugs-for-melanoma/

29 Abrahams E. Personalized medicine in brief. Pers Med Coalit. 2016 Fall;7:2.

5

Toward the Day We Just Call It "Medicine"

Access to Precision Medicine

In visiting the hospital of a small and joyful Caribbean island during a business trip, I was struck by the time warp I had been thrown into. This hospital was built in the 1960s, and few changes had been made since then. On my tour I noticed it was clean and cool from the ocean breezes whipping through the corridors, courtesy of open-air external hallways that let in the weather, which (with exception of hurricane season) was absolutely breathtaking all the time. I also noticed the basic equipment, bare-bones lab facilities, and even the 1960s crisp but outdated style of the nurses passing by. Although the hospital had kept up with basic equipment upgrades, especially around the most general core clinical diagnostics instrumentation and ward bed monitoring, the management explained that anything "special" was going off island, from specialty surgery to cardiac complications to major accidents to cancer care and precision medicine. They just didn't have the resources or the qualified staff to support it. As a native New Englander, I had contemplated a move to the Caribbean for a change of temperature at certain times in my life, but after the hospital tour I immediately snapped back to the reality that the type of care we take for granted in developed countries is just a dream in a place that is magical in so many other ways.

Looking around the world, and as you have read about the regions described throughout this book, the problem of insufficient access to *any* medical care is a major concern, never mind precision medicine. Over one billion people—nearly 15% of the world's population—lack access to effective and affordable healthcare [1]. These one billion people are more commonly found in developing countries. Low- and middle-income countries bear 93% of the world's illness burden—while representing just 11% of global health spending, and less than 20% of world income [2]. Further, there are fewer physicians to deliver care in developing countries. The physician-to-patient ratio in the United States has been increasing or stable over the last several decades because of the stable total number of physicians [3]. But even here in the United States, there are only 3.7 oncologists per 100,000 people [4]. When you think of how many

Personalizing Precision Medicine: A Global Voyage from Vision to Reality, First Edition.
Kristin Ciriello Pothier.
© 2017 John Wiley & Sons, Inc. Published 2017 by John Wiley & Sons, Inc.

people currently have, or even more so, will need to be assessed for cancer over the next decades due to our growing populations, our aging populations, and our inability to significantly reduce lifestyle changes to decrease these numbers, this number of oncologists is not nearly enough. However, the shortage in the United States is minimal compared with what we are seeing in other regions of the world, which I will talk about further by country in the next chapters. When you layer on the complexities of delivering personalized care on a worldwide scale, we have the ultimate access challenge.

There are multiple access hurdles that specifically limit the delivery of precision medicine globally. Four key access issues—preventative care, imaging, *in vitro* diagnostics (IVD), and treatments—are pressing concerns that need to be addressed in order to standardize the access to precision medicine and the delivery of care globally (Figure 5.1).

Before exploring each of these four elements, I want to address the obvious. There are many major global healthcare access concerns—but these do not focus directly on the delivery of precision medicine. Yes, more than one billion people lack access to safe drinking water [5]. Yes, less than 9% of China's urban population lives in a city that meets the national standard for air quality [6]. Yes, less than 30% of the world's population does not have access to sanitation facilities, including flush toilets and covered latrines [7]. Yes, many developing countries lack the roads and transportation infrastructure to deliver adequate care to rural regions. And yes, reimbursement and payment for drugs is a major global concern, but I will address that in the next chapter. Here I want to focus on precision medicine—the actual *access and delivery* of the right treatment to the right patient at the right time. So let us begin.

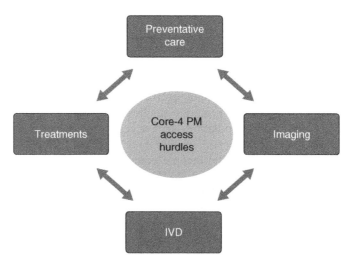

Figure 5.1 The four hurdles to precision care access.

Access to Preventative Care

Polio. Measles. Smallpox. What do all of these diseases have in common? Drugs, called vaccines, have been developed to prevent people from contracting such diseases. Vaccines work by introducing a slightly modified version of a virus or bacteria to the blood stream (in these cases the virus or bacteria will be called a "pathogen"). This process activates the immune system, which is able to use the harmless version of the pathogen to fight off future, harmful versions of the pathogens that might enter the bloodstream later in the person's lifetime.

If these diseases have vaccines that prevent the transmission of disease to patients, then why were there more than 40,000 confirmed cases of measles in Africa in 2015? [8] Why were there over 300 cases of polio in Pakistan in 2015, despite the fact that polio cases worldwide have declined by more than 99% since the 1980s? [9]

It is because people, mostly children, in resource-poor regions of the globe lack access to preventative medicine, especially vaccines. Take pneumococcal vaccine, for instance, which protects children against deadly diseases like sepsis (blood infection), bacterial meningitis, and pneumonia. Globally, pneumococcal disease is responsible for the deaths of more than one million children per year [10]. While companies have developed vaccinations for the disease that cost roughly $50 per child, children and families in many developing nations are simply unable to afford the luxury of preventative care. While reducing the price of vaccinations in resource-poor countries may sound simple, vaccines are "advanced and complex, and require significant upfront capital investment to make and supply," according to a spokesperson from a major vaccine manufacturer [11]. And many leading vaccines are provided to resource-poor countries at steep discounts, at prices as much as 90% lower than in developed regions, but even then the opportunity for widespread vaccination and adoption is low [11]. Bringing low-cost vaccines to resource-poor regions requires a delicate balance of support for research and development, value creation for investors, and charitable giving by the foundation divisions of pharmaceutical companies, which cannot always be executed by offering vaccines globally at close to no cost.

Beyond the financial component of vaccine access, there are additional challenges for preventative medicine in resource-poor countries in the developing world. With regard to vaccines, limited infrastructure and inadequate storage facilities make it difficult to preserve vaccines that require refrigeration in order to maintain the effectiveness of the medication [12]. An even bigger issue is the inability of patients in emerging markets to access preventative care in general. It has been widely proven that citizens in countries taking advantage of preventative care help to control common chronic diseases and lifestyle-related conditions [13]. However—beyond the management of chronic diseases—regular preventative care and engagement with primary care physicians also leads to earlier diagnosis of communicable (i.e., infectious diseases) and noncommunicable (i.e., cancer) conditions. Further, preventative medicine is often delivered through family medicine or primary care physicians, and it has been proven that a good primary care experience is associated with improved self-rating, improved mental health, and reductions in disparities between more- and less-disadvantaged communities in ratings of overall health [14].

Access to *In Vitro* Diagnostics (IVD)

The value of diagnostics has been proven time and time again, from understanding risk, to early detection of diseases, to determining the correct course of action for a patient. However, the positive results associated with the use of diagnostic tests can only be achieved if the patient has the ability to access these technologies. Access to IVD testing in emerging markets is a major limitation to the widespread adoption of precision medicine, and it is not limited to one disease or one country—it is a widespread limitation with many significant barriers to overcome.

In many countries and regions where HIV/AIDS has a higher likelihood of affecting patients, notably African countries, newborns are screened for the potentially life-threatening disease. In the United States, over 90% of HIV-infected women knew of their HIV status before delivery [15]. This is critical, considering that about half of all babies born with HIV will die by the age of 2 years without proper treatment, and it is impossible to deliver proper treatment if the presence of the disease is unknown. Now, let's compare this to Africa—where only 42% of children born to women with HIV received a diagnostic test after birth [16]. This means that nearly three out of five newborns will be unable to access potentially lifesaving medications simply because physicians were unaware that the deadly disease had been transmitted.

While diagnostics are critical for screening and early detection of diseases like HIV, they are also widely utilized for disease management. Patients with HIV suffer because the HIV virus continues to use their cells to replicate and spread, resulting in fatigue and susceptibility to other diseases. Advancements in HIV medications have resulted in new classes of medications called antiretroviral therapies that dramatically restrict the virus from replicating within the patient, helping the patient to live a relatively healthy life. In order to understand whether a patient's virus concentration in their blood—also known as viral load—is under control, there are very useful IVD tests called viral load testing to help the patient learn if the medicine they are receiving is effective. This is especially important since the virus can, in some cases, adapt to resist certain medications. Through regular monitoring of HIV viral loads, patients can be directed to alternative medications and continue to be healthy. The problem is that only 25% of people receiving HIV treatments have access to viral load testing [16]. This means that 75% of patients are potentially paying for medications that they are not responding to, unnecessarily burdened with increased costs of therapy and dreadful side effects.

Targeted therapies—medications that are more effective in patients with certain genetic profiles—only work if those patients are identified. Two of our common lung cancer therapies, Gilotrif (afatinib) and Tarceva (erlotinib), have been proven to increase survival and lower costs in non-small cell lung cancer (NSCLC) patients with epidermal growth factor receptor (EGFR) mutations. These patients—who are EGFR positive—represent about 10% of all NSCLC patients, and nearly 50% of patients who had never smoked, so the profile is less rare compared with other genetic alterations [17]. Due to the increased effectiveness of the new drugs, doctors started testing more and more patients to see if they fit the genetic profile to be a high responder to this new medicine. In 2015, testing for EGFR mutations in newly diagnosed NSCLC patients took place in more than 80% of patients in developed countries (the United States, Europe, South Korea, etc.), resulting in better identification of patients likely to respond to this much more effective drug [18]. However, access to EGFR testing varies dramatically by region, and testing rates are especially low in developing countries and emerging markets. China, where NSCLC rates are among the highest globally, has among the lowest testing rates in the world. Just 26% of Chinese hospitals routinely provide EGFR testing services, and as few as 6% of Chinese patients are routinely tested for EGFR mutations [19].

Access to Imaging Diagnostics

Just because a patient has a positive test result, it doesn't mean that the diagnostic journey is complete. For many patients with a disease, particularly cancer, it is critical to learn—and to see—what is happening beneath the surface of the skin and within the body. To deliver appropriate cancer care requires identification, sizing, and monitoring of the size and shape of tumors within the body, which require the use of advanced imaging techniques. These techniques, largely performed through magnetic resonance imaging (MRI) and computed tomography (CT) scans, are well established in the care paradigm for cancer and other conditions in developed countries. The first publications regarding the utilization of MRI testing date back to the 1970s, although the technique was discovered centuries before [20]. Imaging technologies have continued to evolve and offer higher degrees of specificity and sensitivity related to the differential diagnosis of cancer. However, the benefits of this technological revolution can only be realized if—you guessed it—patients have access to imaging.

There are over 3.6 billion diagnostic X-ray examinations each year globally [21]; however, unfortunately, nearly two-thirds of the world's population has no access to diagnostic imaging, including X-rays, MRIs, CT scans, PET scans, or ultrasounds. This has catastrophic ramifications in the fields of maternal health, CNS disorders, and especially oncology. Pablo Jiminez, PAHO/WHO Regional Adviser on Radiology and Radiation Protection, explains "Access to diagnostic imaging services has a great impact on public health and can potentially reduce, for example, infant mortality, or increase detection of some types of cancer at an early stage. Unfortunately, current shortages of human resources and obsolete or broken equipment are making it increasingly difficult to provide adequate access and quality in [Latin America]" [22]. Despite the fact that 70–80% of diagnostic problems can be resolved through basic imaging, if patients are unable to access these capabilities due to cost, infrastructure restrictions, or availability of trained radiology technicians, the diagnostic support and potentially lifesaving insights will never be realized.

Access to Treatments

With limited preventative care, limited IVDs, and limited imaging, the number of patients who are even able to receive adequate care in the form of medical interventions (surgeries, psychiatric care, etc.) and medications has been reduced. Unfortunately, the picture doesn't get rosier as the paradigm progresses, with access to therapies that the developing world has proving to be just as challenging, if not more difficult, than access to diagnostics and imaging.

Essential medications, which have been major pillars for care in developed nations, have remained out of reach for much of the global population. One of the primary issues plaguing drug access is linked to drug pricing. For standard HIV care in the United States, a regimen of three antiretroviral therapies would cost greater than $10,000, an impossible amount for an average African HIV patient to afford considering the market for these drugs in Africa represents just 1% of global revenues [23]. In the United States, where insurance companies are responsible for most of the cost of pharmaceuticals, access is (somewhat) easier, but for the predominately out-of-pocket market dynamics in Africa, these medications are impossible to procure.

Beyond communicable diseases, access to medications globally for chronic diseases is of paramount concern as well. Diseases ranging from autoimmune conditions to respiratory diseases, all the way to diabetes are all impacted similarly, where patients lack the ability to take the best medications to treat their conditions. This becomes further magnified when we learn that noncommunicable diseases carry among the highest mortality rate and just about the lowest financial assistance, or disbursement threshold [24]. Thus, while the financial access issues are most prominent for HIV/ AIDs and other communicable diseases, additional therapeutic areas (especially noncommunicable diseases) need to be considered and more widely publicized.

The burden of precision medicine access globally cannot be easily solved, but several stakeholders are making colossal efforts to steer the proverbial ship in the right direction. These stakeholders range from federal-level governments themselves, nonprofit organizations and foundations, and private/ public coalitions of industry participants. Let's focus on some of the global leaders.

In South Africa, where the burden of the HIV/AIDS epidemic is disproportionately realized, the government has launched a massive turn-around initiative to provide universal healthcare to all citizens through the implementation of a national health insurance. While the plan hopes to address several aspects related to PM, some efforts with the highest expected benefits are infrastructure recapitalization and improvement in addition to skills development and accreditation [25].

China is also a leader in government-led initiatives to improve access to precision medicine. In 2013, the Ministry of Health and the National Population and Family Planning Commissions merged in an effort to refine the rational use of medicines, including the selection of an essential medicines list. This initiative will enable rural areas to utilize essential medicines with zero markup, allowing for patients in impoverished rural China to access medications that more affluent urban citizens can utilize. Further developments will focus on the rationalization of medical services to impact drug pricing through future policy decisions [25].

Nonprofit groups have also placed a high degree of emphasis on improving access to healthcare, resulting in the higher adoption of precision medicine. Two of the most influential groups at improving access to healthcare have been the Bill and Melinda Gates Foundation (BMGF) and the Clinton Health Access Initiative (CHAI).

BMGF, founded in 2000, is currently the largest transparently operated private foundation in the world. Its endowment of over $40 billion is utilized largely to improve and equalize access to education, healthcare, and technology [26]. BMGF has funded hundreds of initiatives over the last decade, including various efforts to improve access to precision medicine. BMGF has pledged over $2.5 billion to the initiative, but additional investments have been made to bring new therapeutics to market in developing countries where access to standard drugs is impossible [27]. In June of 2015, BMGF earmarked $25 million to support a clinical trial comparing HIV incidence and contraceptive benefits in women using three family planning methods in four sub-Saharan

African countries [28]. The largest of its healthcare initiatives, announced in January 2016, was a pledge of $1.6 billion to the GAVI initiative to "save children's lives and protect people's health by increasing equitable use of vaccines in lower income countries."

Tuberculosis (TB) has also been a hallmark of BMGF investments, as evidenced by a $25 million investment in late 2014 to support a Phase III registrational trial for TB and multidrug-resistant TB [29]. BMGF continues to operate as the foundation by which all other foundations are compared, making dozens of calculated investments to improve access to medications each year.

Beyond BMGF, the CHAI is another industry leader, focused on working with governments and industry players globally to fundamentally change the economics of global health in order to improve access to patient care and precision medicine. CHAI's Access Program focuses on both diagnostics and therapeutics. For diagnostics, it aims to enable access to products used to better diagnose communicable diseases like HIV and TB by working with manufacturers to negotiate heavily reduced pricing. The CHAI helped to structure contracts with HIV viral loading manufacturers, to better target and control HIV/AIDS, and to reduce the cost of tests from roughly $60 to less than $10, which will save more than $150 million by 2019 [30]. For therapeutics, the focus remains in HIV/AIDS and TB treatments, but specifically for children. A partnership with UNITAID, an international consortium on HIV/AIDS and TB, increased the number of children receiving treatment—who would otherwise have been facing a harsh prognosis—from 15,000 to 600,000 in 34 countries from 2004 to 2015 [31].

Nonprofits like CHAI are helping to negotiate with manufacturers, but manufacturers themselves—and coalitions of manufacturers and industry participants—are also working independently to improve access to precision medicine. In 2015, the Diagnostic Access Initiative (DAI) launched a 90-90-90 initiative, which aims to ensure that before 2020, 90% of patients will know their HIV status, 90% of diagnosed patients will be receiving treatments, and 90% of those treated will have a suppressed viral load. This initiative is a product of a major consortium including powerhouses like UNAIDS, WHO, CHAI, PEPFAR, CDC, UNICEF, and others. On the therapeutics side of the coin, Big Pharma powerhouses continue to team up in an effort to provide affordable medications to the developing world. One prominent example has been a partnership between five UN organizations (UNFPA, UNICEF, WHO, World Bank, and UNAIDS) with five Big Pharma representatives to work together to increase access to HIV/AIDS care and treatment in developing countries. By teaming up instead of creating separate agreements, the impact of the pharma coalition is stronger and more widespread, and helps to deliver a stronger punch to the global epidemic.

Did You Know?

Infectious Disease Has Precision Medicine Too

In this chapter, I spend time talking about noncancer diseases to illustrate the importance of access and also to underscore that precision medicine does have roots in noncancer diseases as well, including infectious disease. Infectious disease is unique compared with other areas of disease in the uncertainty and diversity of pathogens that emerge and pose threats to human health and safety. The goal of infection treatment is straightforward: identify and eliminate the "invaders," that is, pathogenic bacteria or viruses, as quickly as possible while not doing harm to the patient.

The development of personalized infectious disease treatments has historically been a reactive one, where innovative treatment evolved out of refining existing approaches that had suboptimal outcomes. One of the earliest applications of precision medicine to modern infectious disease treatment was with the use of the antiviral Ziagen (abacavir) for the treatment of HIV in the early 2000s. Ziagen is an effective and commonly used antiviral for HIV patients, though was found to cause a potentially fatal hypersensitivity reaction in select patients. During treatment, approximately 5% of patients develop an adverse reaction resulting in symptoms such as fever, rash, nausea, and vomiting.

Historically the hypersensitivity reaction could only be identified by clinical diagnosis, in which patients were made aware of the reaction *after* Ziagen had already been administered. Given the effectiveness of Ziagen in treating the majority of patients with whom it was prescribed, there was a need to personalize its use to avoid prescribing it to patients who were likely to develop the hypersensitivity reaction. To do this, one of the hurdles was in understanding what was responsible for this reaction: the virus itself or the patient. In 2002, researchers identified a genetic link between the hypersensitivity reaction observed and a specific variant of a particular family of genes in patients: the class I human leukocyte antigens (HLAs). Patients testing positive for the HLA-B*57:01 variant of the gene had a 60% probability of developing the potentially fatal reaction, while patients testing negative for the variant did not develop the hypersensitivity reaction at all [32, 33].

In the spirit of the first modern application of precision medicine to infectious disease, HIV researchers continued to explore the genetic mechanism of the virus and its interaction with patients in hopes of further advancing personalized HIV therapies. Another innovation came to HIV patients in 2007 with the development of Pfizer's Selzentry (maraviroc), which represented a new class of precision HIV antiretrovirals. Selzentry functions as an entry inhibitor, binding to a critical receptor found in human cells that prevents the HIV virus from entering. Selzentry was unique in that it was the first drug to work by interacting with human cells rather than the virus itself. In order for HIV to infect human cells, the virus may enter through one of two pathways: CCR5 or CXCR4 (think of these as two separate entrances to a house). Selzentry is the first drug to selectively block CCR5, protecting patients from HIV who are positive for CCR5 at a given point in time.

One major precision medicine access issue plaguing developing countries is the availability of medications, often times to the patients who are most in lifesaving need. For patients who have exhausted all of their treatment options and are at the end of the established treatment paradigm, often the only options are palliative care (to manage symptoms) and hospice (to prepare for their death). However, a third option could potentially be in place for these patients, and that is for them to receive experimental drugs that are in development. There is no guarantee that these drugs will work, as they have not been approved and are not available for purchase through traditional channels, but these clinical trials could be the only option that could save a patient's life.

If there is even a possibility that these therapies could save a patient's life when they have no alternative options, then what's the issue in providing them? There are several hurdles. First, the patient needs to be educated to know a trial is available. Next, the patient needs to meet the criteria—which are extremely stringent—to get into the trial. Finally, there is often difficulty being able to stay in the trial, which could be in a completely different state, or area of the world, than a patient's home and support system.

A solution to these concerns that has been gradually increasing in uptake is the initiation of the expanded access programs (EAPs), which will deliver experimental drugs to patients on a compassionate-use basis. Through this program, patients will not be enrolled in clinical trials, so the drug manufacturer will still be able to preserve the highest probability of success in establishing that the drug is effective. The patients that are receiving drugs through the EAP are carefully observed by physicians to watch for side effects and adverse events, since the drugs themselves have not been proven to be either safe or effective yet, and compassionate-use patients are not under the same care and scrutiny of the patients enrolled in the actual clinical trial [34]. However, for many patients with no other options, accessing potentially lifesaving drugs through this channel is the only solution.

Beyond connecting the most sick and in-need patients to drugs just out of their reach, there is one other primary precision medicine access hurdle in place in developed countries: the standardization of care between rural care settings (often community care) and premier, urban academic medical centers. US citizens who live in rural settings are more reliant on government-funded health programs and suffer a higher incidence of chronic diseases, such as hypertension or diabetes [35]. In addition to the increased disease burden, these patients also have less access to primary care physicians to aid in preventative care and early detection or intervention of disease.

As discussed throughout this book, precision medicine is a product of innovative therapies, tools, and thought processes. However, specialized and high-intensity care is difficult to access in rural areas. Emergency medical services (EMSs), for example, utilize less sophisticated measures and are forced to travel over longer distances in rural areas—often resulting in the difference between life and death for patients [36]. Beyond EMS, these patients lack access to the latest technology—such as sequencing analyzers, MRI instrumentation, and latest surgical methods. Physicians in these areas often lack the ability to access clinical trial databases and bioinformatics for making complex treatment decisions.

In a Techonomy.org article, an anecdote was shared in a discussion between Sue Desmond-Hellmann (CEO of BMGF) and Marc Benioff, CEO of Salesforce. com, highlighting the limitations of care in rural settings:

> Let me give you a story on that specific topic, which is that I have a friend of mine who has brain cancer, he has a glioblastoma, and he's had surgery twice to reduce the size of the tumor. And on the second time—he's being operated on in a rural hospital. I said to the neurosurgeon—'Okay, well are we going to type this cancer?' So then she basically said she hasn't sequenced her genes, but sequencing cancer is pretty cool. [37]

The standard of care for glioblastoma, given the rarity of the tumors, is not well established today. However, advances in targeted therapies and molecular informatics have empowered patients and physicians to explore novel approaches through genetic testing and genome sequencing. The fact that a neurosurgeon viewed the option as "pretty cool" and not as an informed way to provide the patient with a more efficacious therapy is an unfortunate circumstance that may not have happened at a leading academic medical center in other parts of the world.

The disparity in care between urban and rural settings is being tackled by a variety of sources and stakeholders through a common platform: telemedicine. Telemedicine, by definition, is the use of medical information exchanged from one site to another via electronic communications to improve a patient's clinical health status [38]. Telemedicine as a platform enables an oncologist in rural North Dakota to communicate with a leading MRI interpretation technician in New York, who can then access thousands of patient records to understand disease prognosis and therapeutic options. While patients may not be able to sit in the same room as leading physicians and medical practitioners, telemedicine enables the comfort of face-to-face interactions through video conferencing.

The level of adoption of telemedicine is highly variable today, but select pockets of innovation exist. Northwell Health System has recently implemented a "tele-ICU" for patients requiring a high degree of isolation, reducing the risk of contamination while maintaining the highest degree of patient care and contact [39]. The University of Miami instituted a pilot telemedicine program, which resulted in improved adherence to specialty care appointments from 34 to 94%, which is dramatic especially when paired with the fact that patients adhering to therapy and doctor's visits generate improved health outcomes. Anthem's John Jesser, President of Anthem's LiveHealth Online regarding telemedicine, stated "We call it telehealth now, but in 5 years, we'll just call it health. You can be face-to-face with a board certified doctor 24/7. That's the "Ah-hah" that's creeping across the category" [39].

Further developments in telemedicine will not only help the United States address the deficit of care in rural settings but will also improve care and access to precision medicine in many metropolitan and urban areas.

Conclusion

This chapter introduced the major issues that are hindering the adoption of precision medicine globally. These issues can be as rudimentary as the availability of diagnostic tests for identifying patients with diseases, or the ability to refrigerate vaccines in Africa. They can also be as complex and nuanced as creating early access programs in developed countries to allow extremely sick patients to use experimental drugs when clinical trials are banning their access to them. While precision medicine has great promise, if patients are unable to access it in the United States and abroad, the full potential cannot be realized.

When addressing access issues in the developing world, we should be cognizant of access issues impacting the developed world today. Technological innovation and connectivity are increasing precision medicine access in the United States and other developed countries, particularly in rural areas. Once these access barriers are lowered, the result may be more rapid acceleration of patient access to precision medicine in developing countries, improved outcomes for patients, and rapid market development for commercial entities, even in nations as small as the Caribbean one mentioned earlier in this chapter. As Mara Aspinall, the Executive Chairman of GenePeeks and Founder of the School for Biomedical Diagnostics at Arizona State says, "The ultimate success of Precision Medicine comes when we just call it Medicine." And access is a major key to realizing that success.

References

1 Shah A. Health issues [Internet]. Global Issues; Sep 27, 2014 [cited Dec 2016]. Available from: http://www.globalissues.org/issue/587/health-issues

2 World Health Organization. The world health report 2000: health systems: improving performance [Internet]. World Health Organization; 2000 [cited Dec 2016]. Available from: http://www.who.int/whr/2000/en/

3 Makaroff LA, Green LA, Petterson SM, Bazemore AW. Graham Center policy one-pager: trends in physician supply and population growth [Internet]. American Family Physician; Apr 1, 2013 [cited Dec 2016]. Available from: http://www.aafp.org/afp/2013/0401/od3.html

4 ASCO. Key trends in tracking supply of and demand for oncologists; Mar 2015. Available from: http://www.asco.org/sites/new-www.asco.org/files/content-files/research-and-progress/documents/2015-cancer-care-in-america-report.pdf

5 WHO. Drinking-water fact sheet [Internet]. World Health Organization; Nov 2016 [cited Dec 2016]. Available from: http://www.who.int/mediacentre/factsheets/fs391/en/

6 GBTimes. 9% of China urban population has access to clean air [Internet]. Gbtimes.com; Dec 22, 2015 [cited Dec 2016]. Available from: http://gbtimes.com/china/9-china-urban-population-has-access-clean-air

7 WHO. Sanitation fact sheet [Internet]. World Health Organization; Nov 2016 [cited Dec 2016]. Available from: http://www.who.int/mediacentre/factsheets/fs392/en/

8 WHO. Measles surveillance data [Internet]. World Health Organization; [cited Dec 2016]. Available from: http://www.who.int/immunization/monitoring_surveillance/burden/vpd/surveillance_type/active/measles_monthlydata/en/

9 Hashim A. Pakistan's polio problem and vaccination danger [Internet]. Al Jazeera English; Mar 28, 2015 [cited Dec 2016]. Available from: http://www.aljazeera.com/indepth/features/2015/03/pakistan-polio-problem-vaccination-danger-150328091807399.html

10 Pneumococcal vaccines [Internet]. WHO; Jul 8, 2008 [cited Feb 18, 2016]. Available from: http://archives.who.int/vaccines/en/pneumococcus.shtml

11 GSK response to MSF vaccine report [Internet]. GSK; Apr 23, 2015 [cited Dec 2016]. Available from: https://au.gsk.com/en-au/media/press-releases/2015/gsk-response-to-msf-vaccines-report/

12 Vaccine delivery—strategy overview. Bill & Melinda Gates Foundation; [cited Feb 18, 2016]. Available from: http://www.gatesfoundation.org/What-We-Do/Global-Development/Vaccine-Delivery

13 Learning from U.S. mistakes, emerging markets commit to universal healthcare coverage [Internet]. Yahoo Finance; [cited Feb 18, 2016]. Available from: https://insurancenewsnet.com/oarticle/learning-from-u-s-mistakes-emerging-markets-commit-to-universal-healthcare-coverage

14 Shi L, Starfield B, Politzer R, Regan J. Primary care, self-rated health, and reductions in social disparities in health. Health Serv Res. Jun 2002;37(3):529–50.

15 MMWR Weekly. HIV testing among pregnant women—United States and Canada, 1998–2001 [Internet]. Centers for Disease Control; Nov 15, 2002 [cited Dec 2016]. Available from: https://www.cdc.gov/mmwr/preview/mmwrhtml/mm5145a1.htm

16 Diagnostics Access Initiative to achieve the 90-90-90 treatment target [Internet]. UNAIDS; Apr 22, 2015 [cited Dec 2016]. Available from: http://www.unaids.org/sites/default/files/media_asset/20150422_diagnostics_access_initiative.pdf

17 Genomic testing [Internet]. Memorial Sloan Kettering Cancer Center; [cited Feb 19, 2016]. Available from: https://www.mskcc.org/cancer-care/types/lung/diagnosis/genetic-testing

18 Ray T. Survey shows more NSCLC patients getting EFGR testing, but results often don't inform care [Internet]. Genome Web; Apr 27, 2015 [cited Dec 2016]. Available from: https://www.genomeweb.com/molecular-diagnostics/survey-shows-more-nsclc-patients-getting-egfr-testing-results-often-dont

19 Goozner M. A tale of two countries: lung cancer care in Brazil and China. J Natl Cancer Inst. Nov 7, 2012;104(21):1621–3.

20 Geva T. Magnetic resonance imaging: historical perspective. J Cardiovasc Magn Reson. 2016;8(4):573–80.

21 Seeram E, Brennan PC. Radiation Protection in Diagnostic X-Ray Imaging. Massachusetts: Jones & Bartlett Learning; 2016.

22 PAHO/WHO Knowledge Management and Communication. World radiography day: two-thirds of the world's population has no access to diagnostic imaging [Internet]. Washington, DC: PAHO and WHO; Nov 2012 [cited Dec 2016]. Available from: http://www.paho.org/hq/index.php?option=com_content&view=article&id=7410%3A2012-dia-radiografia-dos-tercios-poblacion-mundial-no-tiene-acceso-diagnostico-imagen&Itemid=1926&lang=en

23 Keppler H. The untold AIDS story: how access to antiretroviral drugs was obstructed in Africa [Internet]. The EJBM Blog; Oct 1, 2013 [cited Dec 2016]. Available from: https://theejbm.wordpress.com/2013/10/01/the-untold-aids-story-how-access-to-antiretroviral-drugs-was-obstructed-in-africa/

24 Mattke S, Haims MC, Ayivi-Guedehoussou N, Gillen EM, Hunter LE, Klautzer L, et al. Improving access to medicines for non-communicable diseases in the developing world [Internet]. Rand Corporation; 200 [cited Dec 2016]. Available from: http://www.rand.org/pubs/occasional_papers/OP349.html

25 Branco de Araujo GT, Stefani SD, Maksimova L, Bhoi N, Hu S, Brand M. Introduction to health systems of BRICS countries [Internet]. International Society for Pharmacoeconomics and Outcomes Research; 2014 [cited Dec 2016]. Available from: https://www.ispor.org/consortiums/asia/Introduction-to-Health-Systems-of-BRICS-Countries.pdf

26 Foundation fact sheet [Internet]. Bill & Melinda Gates Foundation; [cited Dec 2016]. Available from: http://www.gatesfoundation.org/Who-We-Are/General-Information/Foundation-Factsheet

27 Public-private partnerships [Internet]. Gavi, the Vaccine Alliance; [cited Dec 2016]. Available from: http://www.gavi.org/funding/how-gavi-is-funded/public-private-partnerships/

28 Gabelnick H. Bill & Melinda Gates Foundation announces $25 million grant to CONRAD program's Consortium for Industrial Collaboration in Contraceptive Research [Internet]. Bill & Melinda Gates Foundation;

[cited Dec 2016]. Available from: http://www.gatesfoundation.org/Media-Center/Press-Releases/2000/07/Consortium-for-Industrial-Collaboration-in-Contraceptive-Research

29 Fuller J. The Bill & Melinda Gates Foundation announces new global health grants [Internet]. The Bill & Melinda Gates Foundation; [cited Dec 2016]. Available from: http://www.gatesfoundation.org/Media-Center/Press-Releases/2000/03/HealthRelated-Grants

30 Case study: viral load access program [Internet]. Clinton Health Access Initiative; Feb 2015 11 [cited Dec 2016]. Available from: http://www.clintonhealthaccess.org/case-study-vl-access-program/

31 CHAI annual report 2014 [Internet]. Clinton Health Access Initiative; 2015 [cited Dec 2016]. Available from: http://www.clintonhealthaccess.org/content/uploads/2015/08/CHAI-2014-Annual-Report.pdf

32 Mallal S, Phillips E, Carosi G, Molina JM, Workman C, Tomazic J, et al. HLA-B*5701 screening for hypersensitivity to abacavir. N Engl J Med. Feb 7, 2008;358(6):568–79.

33 Mallal S, Nolan D, Witt C, Masel G, Martin AM, Moore C, et al. Association between presence of HLA-B*5701, HLA-DR7, and HLA-DQ3 and hypersensitivity to HIV-1 reverse-transcriptase inhibitor abacavir. Lancet. Mar 2, 2002;359(9308):727–32.

34 Manna H. Contrasting approaches to early access across the globe [Internet]. Pharmafocus; [cited Dec 2016]. Available from: http://idispharma-staging.com/sites/default/files/uploads/Contrasting-approaches-to-early-access-across-the-globe.pdf

35 American Hospital Association. The opportunities and challenges for rural hospitals in an era of health reform. Trendwatch; Apr 2011 [cited May 8, 2017]. Available from: http://www.aha.org/research/reports/tw/11apr-tw-rural.pdf

36 Rural health disparities introduction [Internet]. Rural Health Information Hub; Oct 31, 2014 [cited Dec 2016]. Available from: https://www.ruralhealthinfo.org/topics/rural-health-disparities

37 Benioff & Desmond-Hellmann: advances, opportunities, and challenges [Internet]. Techonomy; Mar 30, 2015 [cited Dec 2016]. Available from: http://techonomy.com/conf/bio15/big-picture/advances-opportunities-and-challenges/

38 About telemedicine [Internet]. American Telemedicine Association; [cited Mar82016].Availablefrom:http://www.americantelemed.org/main/about/telehealth-faqs-

39 Arnold M. Telemedicine's "Ah-Huh" moment [Internet]. Decision Resources Group; Nov 2015 23 [cited Dec 2016]. Available from: https://decisionresourcesgroup.com/drg-blog/digital-innovation/telemedicines-ah-hah-moment/

6

Precision Medicine around the World

Japan

The sea of apple-sized baby heads topped with shocks of spiked black hair, all cozy in their well-wired bassinets, was overwhelming. I was outside a long glass window looking into the nursery of the National Center for Child Health and Development in Tokyo. Founded in 2002, it is the fifth highly specialized medical center and the largest children's hospital in Japan. The 460-bed medical center houses a 20-bed pediatric intensive care unit (PICU), a 40-bed neonatal intensive care unit (NICU), and a continuum of clinical departments and specialties including everything from basic internal medicine to gene therapy—all designed for catering specifically to moms and kids. It also offers outpatient hospital care, including home care for medically complex children, women's health, and reproductive medicine. It is one of the few centers that research and offer regenerative medicine using both human somatic stem cells and human embryonic stem cells, while the use of the latter remains an ethical debate in other areas of the world. On my tour, it was clear the management team was very proud of their impressive center with such devotion to the littlest humans. But, in a country where people over 65 years are predicted to make up roughly 40% of the total population by 2060 [1], why the seemingly disproportionate focus on moms and kids?

Well, that's the point. There aren't enough children. Japan's birth rate had been shrinking throughout the 1990s and bottomed out in 2005 at 1.29 total fertility rate (TFR). The population projection of the National Institute of Population and Social Security Research, based on surveys and the census the Japanese government conducts every 5 years, found that the overall Japanese population peaked in 2010 at 128 million and has been decreasing ever since. According to the institute's projections, by 2040 the population will be 107.2 million—more than 20 million lower than the current level (Figure 6.1) [2, 3]. As mentioned, the fact that so many in that population will be elderly, and that the actual number of workers in the labor force will decline, raises concerns about the sustainability of the pension system and, in some more rural regions, the future *existence* of some towns.

Personalizing Precision Medicine: A Global Voyage from Vision to Reality, First Edition.
Kristin Ciriello Pothier.
© 2017 John Wiley & Sons, Inc. Published 2017 by John Wiley & Sons, Inc.

Japan Facts

Population: 127 million (2015)

GDP, 2015: $4.1 trillion

Total Population: 127 million (2014)

GDP: $4.7T (2014)

Lung Cancer Incidence: 113,530 (2014)

Figure 6.1 Japan's demographic facts.

Map:

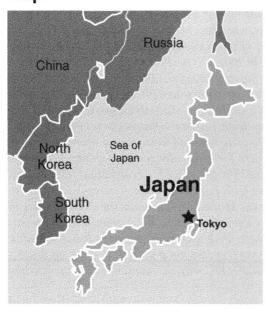

By the time the TFR in Japan had bottomed out, the National Center for Child Health and Development had already opened, with its mission to "... safeguard the health and development of future generations" [4]. This consists of targeting care specifically to expectant mothers to make sure they carry and deliver healthy babies, as well as targeting those babies and the children they become in order to keep them healthy into adulthood. The center seems to have done its part in helping reverse the trend. The birth rate has been slowly but steadily growing, and Japan is now doing better than the other areas of Asia Pacific, including South Korea and Singapore (both 1.19) and Hong Kong (1.12) [5]. To be fair, declining birthrates are being seen all over the world, though Japan continues to be on the lower end (Figure 6.2).

	1960	1970	1980	1990	2000	2010	2011
Australia	3.45	2.86	1.89	1.90	1.76	1.93	1.93
Brazil	6.21	5.02	4.07	2.81	2.36	1.84	1.82
Canada	3.81	2.26	1.74	1.83	1.49	1.63	1.61
European Union	2.58	2.36	1.87	1.66	1.47	1.61	1.58
Japan	2.00	2.14	1.75	1.54	1.36	1.39	1.39
United States	3.65	2.48	1.84	2.08	2.06	1.93	1.89

Figure 6.2 Comparison of birth rates in Australia, Brazil, Canada, the European Union, Japan, and the United States [5].

Even if the birth rate changes considerably over the next years, it still will not change the higher number of elderly people who need to be cared for and the disproportionate number of cancer patients in that elderly population. For example, of the over 100,000 people with lung cancer in Japan, almost 70% of them are patients over the age of 70 years [6]. And in Japan, only 0.7 oncologists exists for every 100,000 people. This is similar to other Asian countries including China (0.6 per 100,000) but in contrast to US and EU regions at 3+ per 100,000 [7]. Only 6.5% of Japan's inhabitants live in rural areas, thus allowing its inhabitants concentrated in major cities easier overall access to those oncologists than in regions like China, where more than 40% of the population lives in rural areas [8].

Therefore, in addition to caring for the newest and (mostly) healthiest in a targeted way, Japan must also care for the oldest and the sickest in a targeted way. Japan has developed a network of oncology care centers across the country, where the majority of those oncologists reside. The top 10 by oncology patient volume include the following [9]:

1) The Cancer Institute Hospital of JFCR
2) Shizuoka Cancer Center
3) National Cancer Center Hospital
4) University of Tokyo Hospital
5) Niigata Cancer Center Hospital
6) National Cancer Center Hospital—East (Chiba)
7) Tokyo Metropolitan Komagome Hospital
8) Kanagawa Cancer Center
9) Osaka Medical Center for Cancer and Cardiovascular Diseases
10) Kyushu University Hospital

While these are the largest, Japan is also home to more regional centers that provide comprehensive services for cancer patients closer to home. For example, after a long, crowded, but impeccably clean subway ride away from Tokyo, I visited Yokohama and Yokohama City University Hospital. Yokohama is the second most populous city in Japan on the coastline of Japan's Pacific Ocean, and Yokohama's port is a major international port [10]. The Yokohama City

University hospital cares for the oldest and sickest patients in a more regional locale. At over 600 beds and as one of the designated Regional Cancer Care hospitals, its clinical labs still house efficient and plentiful equipment for basic anatomical and clinical pathology for a patient. In going through every corner of its clinical labs with my lively and efficient (I think) Japanese interpreter by my side, I was introduced to the inner workings of their lab structure, which, while designed for best-in-class cancer care, had one notable absence: a specialty molecular lab. Like many of the other hospitals across Japan, the most complex testing for precision medicine staging is sent out to one of approximately three major Japanese commercial clinical labs: SRL, Inc.; BML, Inc.; and LSI Medience. All of these labs have in-house IHC, FISH, RT-PCR, and NGS technologies to provide the newest and most innovative precision medicine tests. While visiting those labs, the managers compared and contrasted the merits of their process in Japan versus other countries. Unlike other countries that have tens to hundreds of labs handling precision specialty samples, Japan's ability to have those samples flow through a smaller number of labs allows for much more quality control and much less subjectivity, regardless of whether the tests being done are from a kit or are LDTs.

This is important because, as of this writing, most cancer patients in Japan will get biomarker testing at diagnosis, from EGFR to ALK to BRAF (this is also the typical sequence in Japan) in order to determine their eligibility for targeted therapy and/or clinical trials. On the flip side, Japan faces several challenges that may lessen the growth of access to genetic testing as it moves forward. Nationally, government attitude toward funding of genomic research tends to be conservative. This concern may be due to the undefined ethical territory of genetic testing. Some experts in the field have gone so far as to suggest that Japan's hesitation puts it behind the United States, EU, and China in genomic research and development [11]. As it stands, liquid biopsy and NGS are not yet commonly used in clinical practice today. Over the last few years of discussions with Japanese oncologists, most have estimated that it will take 3–4 years before NGS is used in standard clinical practice, which is longer than the 1–2 years estimated for EU5 countries. Despite this view, highlights are still forthcoming, with NGS companies still pushing to place instruments in the major commercial labs for clinical use and Toshiba rolling out the "Japonica array" genotyping service to support faster advanced genotyping of Japanese individuals for research use [12]. The Japonica Array is a population-specific SNP array developed with over 1000 Japanese individuals, allowing for more advanced genetic population studies [13].

On the reimbursement front, one must begin with an understanding of the universal pubic health insurance system (PHIS), Japan's nationally mandated insurance provider. All PHIS plans offer the same benefit packages covering hospital, primary, and specialist ambulatory and mental healthcare, along with approved prescription drugs, home care services by medical institutions, hospice care, physiotherapy, and most dental care. Preventive measures, specifically including screening, health education, and counseling, are also covered by health insurance plans, while cancer screenings are delivered by municipalities. Meanwhile, private insurance plays a minor role in filling coverage gaps. Thus, testing for genetic alterations is typically funded by the national health system [14]. For such tests, patient co-pays vary from 10 to 20% for those over 70 years of age to 30% for those who are 70 years old or younger. Interestingly, patients often do not pay out-of-pocket expenses when testing for genetic alterations in the absence of approved therapies. Instead, government funding often picks up the tab. Health insurance and patient co-pays are expected to continue to funding biomarker testing and drug costs in the future. Upcoming challenges around reimbursement may not be dire but could alter the current reimbursement system. As new targeted therapies are approved and patients continue to expect to pay the same co-pay rates, physicians mention that reimbursement plans may be increasingly difficult to sustain.

Japan's future with precision medicine will also be shaped in part by the next phases of its genomic medicine project. Its goal is two-pronged: to promote the clinical application of genomic research findings and to strengthen the national genomic research infrastructure. Until 2020, the project will focus on the latter piece by expanding on the legacy work done by BioBank Japan. The second and third components of the research infrastructure are the National Center Biobank Network and major academic centers. A network of six disease-focused biobanks (cancer, CNS, Cardiovascular geriatric, infectious disease/metabolic/autoimmune, and pediatric), the National Center Biobank Network gathers detailed clinic-pathological data that will feed into drug development and personalized medicine. At the same time, major academic centers, such as the Institute of Medical Science of the University of Tokyo (IMSUT) hospital, will work on highly secured supercomputer systems for clinical sequencing and the development of biomedical big data. Together, these three components will carry the genome medicine project into the clinical application phase, lasting from 2020 to 2030. During this decade, two centers will be established to operate on the global forefront of precision medicine. First, the creation of a Central Genome Center will coordinate large-scale genomic research, manage the research infrastructure from phase one, and maintain quality control of samples and information. Second, the development of Medical Genome Center will explore clinical applications of collected genomic information with both academia and industry. The focus of the center will be on the optimization of predictive diagnostics, drug response, and preventative healthcare in therapeutic areas from lifestyle disease (e.g., stroke, diabetes) to cancer. The genomic medicine project, as well as all other funding in the precision medicine space, will be overseen by the newly launched Japan Agency for Medical Research and Development— Japan's version of the NIH. Created in May of 2015, the Japan Agency for

Medical Research and Development wields an initial budget of approximately $1.2 billion to help streamline research funding [11].

Japan has a specific set of challenges that are unique compared with others in the world: its lagging birthrate and its ballooning elderly population in a relatively small ecosystem. However, when it comes to precision medicine, Japan has targeted those two populations that need it the most. If the genome medicine project comes to fruition, Japan will likely emerge with a balanced precision medicine program for all of its residents.

References

1 Dominguez G. Impact of Japan's shrinking population "already palpable" [Internet]. Deutsche Welle; Jun 1, 2015 [cited Dec 6, 2016]. Available form: http://www.dw.com/en/
impact-of-japans-shrinking-population-already-palpable/a-18172873

2 Can Japan boost its low birthrate? [Internet]. Nippon.com; Dec 25, 2014 [cited Dec 6, 2016]. Available from: http://www.nippon.com/en/features/h00089/

3 Statistics and other data [Internet]. Ministry of Health, Labour and Welfare; [cited Dec 6, 2016]. Available from: http://www.mhlw.go.jp/english/database/report.html

4 Kato H. Greetings from the Director of the National Medical Center for Children and Mothers [Internet]. National Center for Child Health and Development; [cited Dec 6, 2016]. Available from: https://www.ncchd.go.jp/en/hospital/about/greeting.html

5 World Bank open data 2013 [Internet]. The World Bank; [cited Dec 6, 2016]. Available from: http://data.worldbank.org/

6 Cancer statistics in Japan – 2014 [Internet]. Foundation for Promotion of Cancer Research; Mar 2015 [cited Dec 6, 2016]. Available from: http://ganjoho.jp/data/reg_stat/statistics/brochure/2014/cancer_statistics_2014.pdf

7 Takiguchi Y, Sekine I, Iwasawa S, Kurimoto R, Sakaida E, Tamura K. Current status of medical oncology in Japan—reality gleaned from a questionnaire sent to designated cancer care hospitals. Jpn J Clin Oncol [Internet]. Dec 2014;44(7):632–40. Available from: https://www.ncbi.nlm.nih.gov/pubmed/24821975

8 Graphiq. China vs. Japan—country facts comparison [Internet]. Find the Data; [cited Mar 13, 2017]. Available from: http://country-facts.findthedata.com/compare/12-82/China-vs-Japan

9 Ranking of cancer hospitals by number of patients discharged [Internet]. Japan Hospital Intelligence Agency Hospital Rankings; [cited Dec 6, 2016]. Available from: http://hospia.jp/Home/Maladylist?mdata=m100

10 Yokohama official visitors' guide [Internet]; [cited Dec 6, 2016]. Available from: http://www.yokohamajapan.com

11 Japan has its own version of NIH [Internet]. Genome Web; May 8, 2015 [cited May2016].Availablefrom:https://www.genomeweb.com/scan/japan-has-its-own-version-nih

12 Toshiba Corporation launches genotyping service using Japonica Array™, a Japanese population genotyping array [Internet]. WebWire; Dec 1, 2014 [cited May 2016]. Available from: http://www.webwire.com/ViewPressRel.asp?aId=193445

13 Kawai Y, Mimori T, Kojima K, Nariai N, Danjoh I, Saito R, et al. Japonica array: improved genotype imputation by designing a population-specific SNP array with 1070 Japanese individuals. J Hum Genet. Oct 2015;60(1):581–7.

14 Mossialos E, Wenzl M, Osborn R, Sarnak D. 2015 international profiles of health care systems [Internet]. The Commonwealth Fund; Jan 2016 [cited May, 2016]. Available from: http://www.commonwealthfund.org/~/media/files/publications/fund-report/2016/jan/1857_mossialos_intl_profiles_2015_v7.pdf

7

Shifting Rules

Regulation and Reimbursement in Precision Medicine

There is a certain place in any book where a chapter needs to be written about the obvious. We have talked about the foundation of precision medicine, the patient journey, and the overall access to the diagnostics and the therapeutics at hand. However, these drugs and diagnostics, in most countries, are officially regulated by governing bodies and are paid for or "reimbursed" in many countries by multiple governing bodies, thus presenting specific stage-gates and challenges for patients to gain access to novel diagnostics and therapeutics. These two topics, "regulation" and "reimbursement," are two of the most volatile and dense topics in precision medicine. They are highly changeable, sometimes contradictory, and usually complex. Mara Aspinall, Executive Chairman of GenePeeks and Founder of the School of Biomedical Diagnostics at Arizona State University, sums the two topics nicely. "Our regulatory and reimbursement system must change. We aggressively regulate diagnostic pricing and lightly regulate their approval—but for drugs—we aggressively regulate their approval and lightly regulate their pricing."

Unfortunately, this has been the norm of both regulation and reimbursement for the last several decades, encompassing all of precision medicine. But it is crucial to understand at least the baseline we are faced with. In this book, I will cover basics on the US market to give you a start and provide references for you to keep up and obtain more information as the areas change over time.

Drug Regulation in the United States

Imagine a world where food and drug manufacturers do not need to list their products' ingredients and little is known about the impact of various ingredients on human health. "Therapies" containing opium, heroin, and cocaine can be sold without restriction. Pregnant women can take the sedative thalidomide without being made aware that thalidomide is associated with birth defects.

Personalizing Precision Medicine: A Global Voyage from Vision to Reality, First Edition.
Kristin Ciriello Pothier.
© 2017 John Wiley & Sons, Inc. Published 2017 by John Wiley & Sons, Inc.

While these unfortunate incidents sound implausible today, all of these incidents and more were commonplace just a few decades ago in the United States [1].

Fortunately, legislation has since been implemented to protect consumers from unsafe food and drugs. For example, the Federal Food, Drug, and Cosmetic Act of 1938 required drug manufacturers to provide safety data on new products. In addition, a regulatory body, the US Food and Drug Administration (FDA), is in place to evaluate and ensure the safety and efficacy of food and drugs, including targeted therapies. The FDA's responsibility to protect consumers is a formidable task, especially because almost all drugs, even efficacious ones, have side effects. Thus, it is up to the FDA to assess whether a drug should be approved based on trade-offs across efficacy, safety, and other factors. While the FDA has approved drugs that save lives and rejected many drugs that are deemed unsafe or inefficacious, it is not uncommon to see headlines like "The FDA's Deadly Track Record" in the news [2]. This has encouraged the FDA to create a thorough and lengthy process for drugs to come to market, with the median targeted therapy taking approximately 7.5 years to go through the FDA approval process [3, 4].

Targeted Therapy Approval Process and Challenges

The FDA's drug approval process is a complex, multi-step process. A drug will undergo multiple applications, preclinical trials, clinical trials, inspections, and even post-marketing review after a drug receives approval (Figure 7.1).

For the most part, precision medicine targeted therapies undergo the same approval process that nontargeted therapies undergo. However, precision medicine faces a few unique regulatory challenges. The first challenge is that targeted therapies, of course, must show benefit in specific populations, so clinical trials must include patients from that specific population. For example, a disease like cystic fibrosis affects just 70,000 people in the whole world (0.001%, or 1 out of every 100,000 people!) [6]. Furthermore, Kalydeco (explained in an earlier chapter) benefits only the 4–5% of cystic fibrosis patients who have a G551D mutation in the CFTR gene. Thus, Kalydeco's clinical trials required scientists to find and recruit patients from a pool of only 3,000 patients in the world who have both cystic fibrosis and a G551D mutation.

Furthermore, targeted therapies are linked to biomarkers, which adds another layer of complexity. During the development of Iressa, the targeted therapy in lung cancer that is associated with an EGFR mutation, researchers measured EGFR via three methods: EGFR protein expression, EGFR gene copy number, and EGFR mutation. While it is now understood that EGFR mutation is the strongest predictor in identifying patients that will benefit from Iressa, this was not known during the initial development of Iressa. This predictive

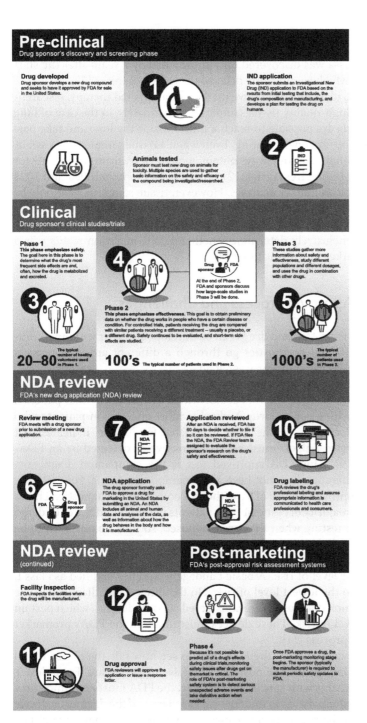

Figure 7.1 FDA drug approval process. Source: Adapted from Center for Drug Evaluation and Research [5].

biomarker for Iressa was discovered only approximately 7 years after the start of clinical trials, which significantly delayed the access of Iressa [7].

Additionally, to identify a predictive marker for a targeted therapy, researchers must collect high-quality tissue samples. Unfortunately, in the case of lung cancer, it is not always possible to reach a tumor inside the lung and take part of the tumor tissue for genetic testing. Even when a tissue sample is accessible, medical professionals cannot always obtain enough DNA from the sample for genetic analyses. In one Iressa trial, 1217 participants were enrolled and 1038 agreed to undergo biomarker analyses. However, tissue samples could only be obtained from 683 patients, of which only 437 were evaluable for EGFR mutations. Thus, many Iressa studies had insufficient tissue samples, and researchers could not conclusively determine whether EGFR mutation status was a predictive biomarker for Iressa at first [8].

Diagnostics Approval Process and Challenges

In precision medicine, targeted drugs can be developed with or without a companion diagnostic, and diagnostics go through a different approval process with FDA, if they go through a process at all.

As discussed throughout this book, companion diagnostics are the precision medicine gateways that help doctors decide which treatments to offer patients. For example, the lung cancer drug Xalkori is most effective in patients who have ALK translocations, so the FDA drug label for Xalkori indicates that patients who receive Xalkori must be "ALK-positive as detected by an FDA-approved test" [9]. This FDA-approved test is the companion diagnostic. Other targeted therapies may not have a companion diagnostic. For instance, Kalydeco was approved by the FDA without a companion diagnostic. This is because the vast majority of cystic fibrosis patients in the United States undergo genetic mutation testing when they are diagnosed, so they do not need to be retested to receive Kalydeco [10].

Diagnostics and other devices, unlike drugs, are classified by the FDA according to their level of risk. A basic bandage would be classified as class I, which pertains to the lowest level of risk. On the other hand, diagnostics like Xalkori's companion diagnostics are considered class III devices, which pertain to the highest level of risk. These devices must undergo the FDA's premarket approval (PMA) process [11].

In the PMA review process, the FDA tests to see how safe the device is and how serious the impact of false positives and false negatives are. For a device that diagnoses patients with cancer, for example, the FDA must determine how costly it would be for patients to undergo cancer therapy if the diagnostic device identified them as having a specific cancer biomarker when they were in fact cancer-free, which is called a false positive. If the cost is high for that false positive diagnosis or it occurs frequently, the FDA will send that device back for further modifications. For most PMAs, the device's sponsors will identify outcome end points, which are events or outcomes that can be measured objectively to determine whether the diagnostic being studied is beneficial. The device's sponsors must also establish the device's performance in relation to those end points using clinical studies. If a device successfully undergoes this process, the device is approved for manufacture and sale in the United States.

Unfortunately, the regulatory pathway for diagnostics can be unclear, and the FDA acknowledges that "there are few performance standards on which to ensure that regulatory decisions are based on clearly defined scientific parameters" [11]. Recent concepts for regulatory reforms have yet to be fully aligned, and Amy Miller, Ph.D., at the Personalized Medicine Coalition (PMC) states that "for many years, various groups contemplated legislative alternatives to both the current device regulation process and FDA's regulation of LDTs. Comparison and discussion of those ideas has recently turned toward areas of commonality, not differences," which include a tiered approach to regulation, a focus on higher-risk tests, and the poor fit of the device regulatory framework for diagnostics [12].

There is also no minimum accepted safety standard across device types [11]. Some argue that this relatively weak regulatory bar for novel diagnostics has put more pressure on payers to serve as gatekeeper for these technologies. Drugs that are strong enough to pass the FDA approval process are usually reimbursed by most payers. However, because the diagnostics review process is less defined compared with the drug review process, payers are uncertain whether or not to reimburse diagnostics that pass through the diagnostic review process. This creates an environment of uncertainty and inconsistency when insurance companies decide whether or not to reimburse diagnostics, which we will discuss further in the next section [13].

Improving the Regulatory Pathway in Precision Medicine

Recent advances may allow us to overcome some of the regulatory challenges observed in precision medicine. Biomarkers for the purposes of precision medicine include ALK, BRAF, KRAS, EGFR, HER2, BRCA, PD-L1, and ROS1. From a pharmaceutical manufacturer standpoint, AstraZenenca's regulatory initiatives in precision medicine offer a case study on how diagnostics and drugs that are used in personalized medicine can be developed more quickly.

It took about 7 years to identify EGFR as a predictive biomarker for Iressa, and EGFR is much better understood today. AstraZeneca, who gained FDA approval for the targeted lung cancer drug Tagrisso in November 2015, collaborated with Roche early on to develop the cobas® EGFR Mutation Test v2 as a companion diagnostic. Unlike Iressa, Tagrisso was approved by the FDA in just 2.5 years after clinical trials began due in part to its association with a strong predictive biomarker and an established biomarker test [14].[1]

Furthermore, although tissue collection remains challenging, several techniques are being developed that will improve the ability of researchers to run more genetic tests even with limited tissue availability. AstraZeneca and Qiagen have co-developed the Qiagen therascreen EGFR Plasma RGQ PCR Kit, which analyzes EGFR mutation status from blood [15]. This will enable patients who do not have sufficient tissue samples to be tested for EGFR mutations via blood samples in lieu of tissue. Testing via blood is much less invasive for the patient than tissue biopsies, and diagnostic developers are working to develop liquid biopsy-based tests for a variety of other genetic alterations.

In addition to these technical advances, the Obama administration had taken steps to improve the speed of FDA review. On July 9, 2012, the Federal Government passed the Food and Drug Administration Safety and Innovation Act. Within this act is an amendment to the Food, Drug, and Cosmetic Act to establish a new classification of drugs designated as "breakthrough therapies" [3]. Breakthrough therapies are drugs that are intended to treat a serious or life-threatening disease or condition where the preliminary clinical evidence indicates that the drug may be superior to existing therapies [3]. A drug developer can request that a drug be reviewed as a breakthrough therapy, and, if approved, it will receive expedited communication and feedback from the FDA with the objective of obtaining drug approval more quickly. Of the 19 breakthrough-therapy designations that were granted between 2012 and 2013, 73% were for targeted drugs [4]. For example, AstaZeneca's Tagrisso and Novartis' Zykadia were each granted breakthrough status and approved several months ahead of the FDA's official decision deadline [16].

To further increase the efficiency of the drug approval process, the FDA has offered advice in recent years on the structure of clinical trials, development milestones to target, and data requirements [17]. An analysis by the FDA Center for Drug Evaluation and Research found that drug developers who received advice in meetings with the FDA prior to their IND application had a 5-year shorter average development time, going from 11 to 6 years [3]. Zineh and Woodcock found that when the FDA advised manufacturers at the end of phase 1, the average development time of the drugs dropped by a year and a half [3]. For example, Zykadia received advisory services from the FDA, and as a result, Novartis, which developed Zykadia, was able to ensure that its clinical

---

1 TAGRISSO™ (AZD9291) approved by the US FDA for patients with EGFR T790M mutation-positive metastatic non-small cell lung cancer.

trials satisfied the FDA's safety and efficacy requirements, contributing to Zykadia's approval ahead of schedule [16].

Addressing the technical challenges associated with clinical trials, such as sample collection, allowing promising therapies to be fast-tracked through the regulatory process, and adding FDA advisory services to define FDA expectations early on will allow new advances in precision medicine to reach the market faster. Also, there is a final added complexity in diagnostics in precision medicine. Most diagnostics never go through FDA review at all. These tests are called lab-developed tests (LDTs), and while subject to scrutiny by Clinical Laboratory Improvement Amendment (CLIA), they are able to be done in a specific lab without being reviewed by FDA. In 2014, FDA released a framework for their oversight of LDTs, and organizations across the United States responded to this. In 2016, FDA announced its decision to delay finalization of its guidance on the topic. The uncertainty surrounding the future of the regulatory landscape for LDTs continues to discourage investment in diagnostics [18].

Did You Know?

Precision pain management is a top priority for both payers and physicians

Chronic pain currently has no cure and affects approximately 100 million US adults, with national healthcare costs ranging from $560 to 635 billion annually [19]. Pain is an especially difficult condition to treat, and there is no uniform standard for the treatment of pain. Physicians also have no way of objectively assessing a patient's pain, relying instead on various subjective scales including the 0–10 visual analog scale and patient behavioral assessment scales.

While there are countless therapies, surgeries, and medications that may be used to treat pain, one of the strongest treatments is an opioid prescription. The risks associated with opioid use are well known with the number of prescription opioid overdoses having increased more than 300% in the past 15 years, now surpassing deaths due to overdoses from all illicit drugs (e.g., heroin, cocaine) combined [20]. While doctors can look for warning signs such as family history of drug abuse and mental health issues, their insight to the risk for their patient becoming such a statistic is limited. However, genetic tests have potential to be highly impactful in pain management, through their ability to assist doctors in prescribing the right therapies to manage patients' pain.

The most effective way to combat these issues is through a drug metabolism test. Companies such as Genelex, Harmonyx, and Millennium Health have drug metabolism tests designed to evaluate a patient's response to pain medications. Other companies, such as Genomind, Iverson Genetics, PGXL Laboratories, and Prenetics have drug metabolism tests that more broadly evaluate a patient's response to commonly prescribed drugs, at times including pain medication. Finally, companies like Kailos and Pathway Genomics perform direct-to-consumer genetic testing, including results for an individual's response to pain medications. However, these tests tend to assess a smaller array of genes than do tests administered through care providers. The tests mentioned earlier examine a variety of SNPs. The most common DNA sequence tested for variations is CYP450, which is responsible for coding a family of enzymes that regulate metabolism. CYP450 tests help determine the rate at which an individual metabolizes medications. While all of the different SNPs tested are responsible for the metabolism of different specific medications, some SNPs interact with each other (e.g., CYP3A4 and CYP3A5 work together in drug metabolism), and some drugs can be metabolized by multiple SNPs (e.g., Tramadol is metabolized by CYP3A4 and CYP2D6, so both must be assessed to determine the patient's metabolism of the drug). The goal of tests that assess a variety of different SNPs is to help providers determine the correct drug and dosage for a patient and evaluate the types of risks and responses the patient might have to a medication.

The two major obstacles to the more widespread use of these tests is direct linkage to improved clinical outcomes (more difficult to prove in pain as measures are both objective and subjective) and therefore reimbursement for the tests from payers. However, because chronic pain is such a cost to any healthcare system, and physicians and payers are both keen to address it, this emerging field is one to watch.

Drug and Diagnostic Coverage

Unfortunately, once a drug with or without a diagnostic is approved, access to the drug is no guarantee. In fact, stakeholders indicate that the paramount challenge to precision medicine is reimbursement, not scientific and technical hurdles.

Imagine a patient, Anne, who is diagnosed with metastatic lung cancer. She is young and has never smoked, which are characteristics that are often seen in lung cancer patients who have genetic alterations. Therapies that target certain genetic alterations in lung cancer patients, such as Xalkori and Zykadia, are already approved and could benefit Anne. However, Anne's insurance company may not agree to cover the cost of diagnostic tests. Even if her insurance company does agree to cover the cost of diagnostics tests, Anne may be asked to pay 20–30% of the total cost of the drug, which could amount to as much as $36,000 per year, given that all cancer drugs that were approved by the FDA in 2014 were priced above $120,000 per year of use. As a result, despite having some insurance coverage, Anne may not be able to afford treatment given that the average gross income in the United States is $52,000 per household per year [21]. This chapter will address some of the challenges that Anne, other patients, and medical professionals may encounter when seeking reimbursement for personalized medicine:

> Regulatory bodies around the world now view information on specific responding populations as a critical piece of a submission package. From development through approval, new therapies are relying on the precision medicine information provided by diagnostic testing. It thus remains paradoxical, given the impact of diagnostic information, that diagnostic tests routinely lag therapeutics in investment, adoption and reimbursement. When validated and used properly, diagnostic testing could save much more than their reimbursement rates in finding the right therapy for the patient the first time. So why isn't diagnostic reimbursement valued more for what it can bring to the appropriate use of therapeutics in maximizing care and cost effectiveness?
>
> *– Glenn Miller, President, CDx Vision, LLC*

In the United States, following passage of the Affordable Care Act, the number of insured people increased from 86.7% in 2013 to 89.6% in 2014. The majority of Americans (66.0%) have private health insurance, instead of government coverage, and some have both. Over 50% of the insured population has employer-based insurance, 19.5% has Medicaid coverage, and 16.0% has Medicare insurance. And overall, reimbursement varies based on setting of drug distribution. Retail and mail-order pharmacy of self-administered drugs are reimbursed on a fee-for-service basis by private insurance on Medicare's Part D, inpatient hospital admission drugs are covered by inpatient benefits, and drugs dispensed in clinical practice are covered by a medical benefit.

US payers, which include insurance companies as well as government bodies like Medicare and Medicaid, have formal processes when deciding whether they will pay for the therapies that their members use. For example, most payers have groups of medical professionals, such as a pharmacy and therapeutics (P&T) committee, which meets regularly to establish coverage policies for new drugs. Even before a drug is approved, drug manufacturers often know what type of data is needed to obtain coverage once the drug is approved. This allows biopharmaceutical companies to collect relevant data during their clinical trials and to present this data for coverage decisions so that patients can access new therapies as quickly as possible.

In contrast with the review process for drugs, the payer's review process for diagnostics is more variable. Often, the review process does not begin until physicians begin to demand coverage for novel diagnostics. Once the review process begins, the lack of standardization and transparency makes it difficult to know what evidence diagnostic developers should provide. The lack of standardization leads to coverage inconsistencies, meaning that some patients will be reimbursed for certain diagnostics, while other patients will not be reimbursed for the same diagnostics simply because their payer has decided not to cover them.

Even large payers, such as Medicare, have not standardized their approval process for diagnostics. The Center for Medicare and Medicaid Services (CMS) can issue coverage decisions that apply to Medicare patients across the nation. However, CMS typically allows its regional contractors to set local coverage policies [22]. This forces diagnostic developers to secure coverage among many different players, which increases costs for the diagnostic developer and in turn increases the price of the diagnostics.

Because companion diagnostics are tied to specific drugs, they have been successful at gaining more consistent coverage. For example, the FDA label for Gilotrif, a drug that targets non-small cell lung cancer patients with EGFR mutations, requires the use of an FDA-approved diagnostic test. Thus, payers who decide to cover Gilotrif know that patients will need the companion diagnostic in order to receive Gilotrif. This makes the reimbursement of Gilotrif's companion diagnostic, Qiagen's EGFR test, more straightforward.

Coding and Payment

Coding also poses challenges to the reimbursement of targeted diagnostics. Historically, there were three methods of coding: code stacking, miscellaneous coding, and new code creation.

Code stacking used reimbursement codes for existing laboratory analyses to bill for new diagnostics. For example, when the Oncotype Dx test was developed, payers used existing codes for various laboratory analyses, such as extraction of genetic material, to bill for the Oncotype Dx test. Each of these analyses has an existing CPT code, and the CPT codes were stacked or added on top of each other to bill for a procedure like Oncotype Dx. This method allowed diagnostic companies to receive reimbursement faster. While this method accounted for the costs of lab analyses, it did not account for the diagnostic company's additional development costs, such as the cost of proprietary algorithms that are needed for biomarker testing. Furthermore, payers disliked this method as it made it challenging to determine which specific diagnostic test(s) were administered to their members. Code stacking was eliminated in 2013 to simplify the amount of time spent on coding and to more easily track which tests were being used [23, 24].

Another coding method, miscellaneous coding, was historically used to allow manufacturers to establish a miscellaneous code with reimbursement rates determined by the manufacturer. This set the reimbursement rate at a level that represented the value of the test. Payers saw miscellaneous codes as red flags and also disliked that miscellaneous codes did not allow them to track the volume of usage of specific diagnostics. By using miscellaneous codes, diagnostic manufacturers were able to achieve reimbursement levels that were approximately six times higher than that which they would receive through code stacking. For example, Oncotype Dx set reimbursement rates at $3460, more than six times higher than the rate of $546.29 from code stacking. Due to the expansion of coding by the American Medical Association (AMA), this coding system has fallen increasingly out of favor.

The most prevalent coding scheme, historically and currently, is CPT coding, which creates new codes for procedures. Each procedure is assigned a brand new reimbursement code that is unique to the new diagnostic, which can be obtained from the AMA. Payers favor the ability to use unique codes to track the volume of diagnostic usage among their members. On the other hand, obtaining a unique code can take several years and does not guarantee that payers will cover the test. If it is covered, the level of reimbursement may be low. Furthermore, if the test is perceived as investigational, it is unlikely to be reimbursed at all.

The transition away from code stacking and miscellaneous codes in targeted medicine has been beneficial for payers, allowing them to better track the usage of targeted therapies in their system [13]. However, the requirement to make a new reimbursement code for each diagnostic has become a logistical challenge for doctors and for the AMA [25]. For instance, the ICD-10, which provides new diagnosis codes for almost all known diseases, including targeted diagnostics, has increased the number of disease codes fivefold since the last iteration (ICD-9), from 13,000 to 68,000 [25].

The logistical challenges from the increase in diagnostic codes has been especially problematic in targeted medicine because of targeted medicine's increasing importance in the past few years and the similarities between targeted therapies for a single condition [26]. For instance, the EGFR gene mutation for non-small cell lung cancer alone has over twenty targeted therapies that require different codes [27].

According to the AMA ICD-10, this dramatic increase "[has negatively affected] claims submission and most business processes within a physician's practice" [25]. Physicians must spend more time and resources making sure they are billing the right reimbursement codes when there may be dozens of similar codes for similar therapies [25]. It has also increased the burden on the AMA to write comprehensive reimbursement codes in line with the HCPCS, the CMS's procedure guidelines for writing codes [25].

All of this complexity is the price for the ICD-10's superior accuracy and modernization [28]. The ICD-10 identifies new technologies and procedures, which helps clarify the coding of targeted medicine [29]. The ICD-9 did not have the language necessary to describe newer targeted therapies, which was a hurdle during the reimbursement process [29]. Now, the descriptive language serves as more convincing evidence for medical necessity, and the reimbursement rate of targeted therapies will likely rise, although this remains to be seen given that ICD-10 was implemented in the United States in late 2015. This provides some help to offset the difficulties in coding described earlier. Furthermore, the ICD-10 has codes with both symptoms and diagnoses to make codes more descriptive without requiring code stacking. ICD-10 was made with reimbursement coding in mind, while ICD-9 was created and implemented in the 1970s and not suited for today's reimbursement system [29].

These changes with ICD-10 benefit payers most, as they will be able to better view medical procedure trends, analyze risk, and better target treatments to their members [29]. In spite of the logistical hurdles, the US government is willing to accept the complexity of ICD-10 due to its many benefits. While these logistical challenges will increase resources spent on coding, especially in precision medicine, ICD-10 will be widely implemented to improve overall efficiency and modernization of the healthcare market [28].

Did You Know?

Market access and reimbursement for novel diagnostics is a worldwide concern. For example, in 2012, UK National Institute for Health and Care Excellence (NICE) provided guidance that recommended CML drug Tasigna based on its efficacy and discounts provided by its company manufacturer to the UK health system, but did not recommend Sprycel, the competitor drug. In 2016, NICE recommended Sprycel based on further clinical trial results, and a new set of discounts via a patient access program. Indeed, reimbursement for cancer therapeutics is variable and is variable by country. See Figure 7.2 for more details.

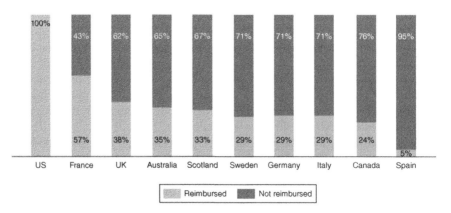

Figure 7.2 Proportion of cancer medicines reimbursed (example countries).

Health Economic Studies

To help justify reimbursement prices that are set by the drug or diagnostic manufacturer, some companies release health economic studies. The purpose of these studies is to show that the use of the drug or diagnostic will lead to cost savings that will more than cover the cost of the drug or diagnostic. For example, the paper "Economic Analysis of Alternative Strategies for Detection of ALK Rearrangements in Non-Small Cell Lung Cancer" shows that the cost for prolonged lung cancer treatment is very high, so determining whether the patient can be helped by targeted ALK treatment is cost effective [30]. The study then assessed various ALK diagnostic tests and determined that the Ventana immunohistochemistry (IHC) test is most cost-effective [30].

Unfortunately, health economic studies have been met with differing reactions in the United States. Some payers find these studies to be helpful when making coverage decisions. Other payers believe that health economic studies, especially when they're sponsored by diagnostic or drug manufacturers, are nothing more than marketing tools.

On the other hand, health economic studies are more widely accepted in Europe because they are commissioned by more objective sources, such as government agencies. For instance, the UK's NICE releases reports on diagnostic treatments called DCDs, and in Germany, diagnostics are reviewed and recommended based on HTA analysis by a laboratory working group (AG Labor) [31, 32]. NICE's commission to safeguard the populace and lower the overall costs of the health system limits bias, and "registered stakeholders" (i.e., people with a financial or medical interest in the treatment) have the opportunity to provide input on these studies [31]. Economic studies have the potential to lower overall healthcare costs by identifying the most cost-effective diagnostics and therapies, and the United States may be able to learn from Europe when executing economic studies that can be credibly used to inform reimbursement decisions.

Many stakeholders believe that a new era of diagnostics reimbursement that makes more sense to the system is around the corner. James McCullough, a partner at Renwick Capital (a life sciences investor) and former diagnostics company CEO, says, "Amazingly, reimbursement and incentives for precision diagnostics have been stuck in an old world system developed before email. Fortunately, change is arriving as payer systems and consumer patients begin see the critical value genetic diagnostic technologies bring our lives. Pricing power will continue to shift into precision diagnostics that can detect disease early and properly direct new therapies to patients who actually benefit. Better pricing means more available diagnostic investment capital across a range of diseases."

Indeed, the regulatory and reimbursement landscape in precision medicine is still evolving and is highly changeable. One of the most up-to-date authorities is the PMC (www.personalizedmedicinecoalition.org), whose leaders frequently publish detailed reports, speak about the challenges, and actively push change. While speaking with longtime colleague Daryl Pritchard, Vice President of Science Policy at PMC, he reflected that strength in action moving forward, saying, "Now is the time to act in order to capitalize on the opportunity of personalized medicine, and the path will be cleared if concrete reimbursement and diagnostic regulatory oversight policies are put in place."

Creating standardized guidelines for regulation and coverage will help ensure that patients have access to precision medicine regardless of what type of medical insurance they have. Encouraging objective health economic studies will allow payers to more easily assess which therapies are most cost effective and ultimately reduce the perceived high cost of healthcare. And a continued question from major stakeholders on whether the current system of treating the diagnostic as part of the drug delivery rather than a true value guiding gateway to the drug delivery remains. "We can deliver comprehensive precision medicine including identification of responder subtypes, drug regimens and dosing management, but this would require a paradigm shift towards value-based compensation, to increase the pressure to Pharma to engage in these concepts," says Thomas Wilckens, leader of the Precision Medicine Group and CEO of InnVentis, a novel precision medicine company. This paradigm shift has yet to come.

References

1 Janssen WF. The story of the laws behind the labels [Internet]. Food and Drug Administration; Jun 1981 [cited Jan 25, 2016]. Available from: https://www.fda.gov/AboutFDA/WhatWeDo/History/Overviews/ucm056044.htm

2 Trowbridge RL, Walker S. The FDA's deadly track record [Internet]. The Wall Street Journal; Aug 14, 2007 [cited Jan 25, 2016]. Available from: https://www.wsj.com/articles/SB118705547735996773

3 Zineh I, Woodcock J. Clinical pharmacology and the catalysis of regulatory science: opportunities for the advancement of drug development and evaluation. Clin Pharmacol Ther. Jun 2013;93(6):515–25.

4 Pacanowski MA, Leptak C, Zineh I. Next generation medicines: past regulatory experience and considerations for the future. Clin Pharmacol Ther. Mar 2014;95(3):247–9.

5 Adapted from: Center for Drug Evaluation and Research. FDA drug approval process infographic [Internet]. U.S. Food & Drug Administration; [cited Jan 25, 2016]. Available from: https://www.fda.gov/Drugs/ResourcesForYou/Consumers/ucm295473.htm

6 Davies JC, Wainwright CE, Canny GJ, Chilvers MA, Howenstine MS, Munck A, et al. Efficacy and safety of ivacaftor in patients aged 6 to 11 years with cystic fibrosis with a G551D mutation. Am J Respir Crit Care Med. Jun 1, 2013;187(11):1219–25.

7 Schlisky RL. Drug approval challenges in the age of personalized cancer treatment. Personalized Medicine. 2011;8(6):633–40.

8 Armour AA, Watkins CL. The challenge of targeting EGFR: experience with gefitinib in nonsmall cell lung cancer. Eur Respir Rev. 2010;19(117):186–96.

9 Food and Drug Administration. FDA approves xalkori with companion diagnostic for a type of late-stage lung cancer [Internet]. U.S. Food & Drug Administration; Aug 23, 2011 [cited Feb 9, 2016]. Available from: http://www.prnewswire.com/news-releases/fda-approves-xalkori-with-companion-diagnostic-for-a-type-of-late-stage-lung-cancer-128484413.html

10 American College of Obstetricians and Gynecologists Committee on Genetics. ACOG Committee opinion no. 486: update on carrier screening for cystic fibrosis. Obstet Gynecol. 2011;117(4):1028–31.

11 Food and Drug Administration. Overview of IVD regulation [Internet]. U.S. Food & Drug Administration; Mar 19, 2015 [cited Jan 2016]. Available from: https://www.fda.gov/MedicalDevices/DeviceRegulationandGuidance/IVDRegulatoryAssistance/ucm123682.htm

12 Miller AM. Preparing for the inevitable? FDA's regulation of LDTs [Internet]. Personalized Medicine Coalition; Mar 3, 2016 [cited Dec 2016]. Available from: https://personalizedmedicine.blog/2016/03/03/preparing-for-inevitable-fdas-regulation-of-ldts/

13 Gustavsen G, Philips K, Pothier K. The reimbursement landscape for novel diagnostics: current impact, real-world impact, and proposed solutions [Internet]. Health Advances; 2010 [cited Apr 10, 2017]. Available from: https://healthadvances.com/admin/resources/noveldiagreimbursement.pdf?

14 TAGRISSO™ (Osimertinib) tablet, for oral use initial U.S. approval [Internet]. U.S. Food & Drug Administration; 2015 [cited Feb 3, 2016]. Available from: http://www.accessdata.fda.gov/drugsatfda_docs/label/2015/208065s000lbl.pdf

15 QIAGEN N.V. QIAGEN companion diagnostic wins FDA approval for use with IRESSA [Internet]. PRNewswire; Jun 13, 2015 [cited Feb 3, 2016]. Available from: http://www.prnewswire.com/news-releases/qiagen-companion-diagnostic-wins-fda-approval-for-use-with-iressa-514677921.html

16 FDA approves Zykadia for late-stage lung cancer [Internet]. U.S. Food & Drug Administration; Apr 29, 2014 [cited Jan 25, 2016]. Available from: http://www.pharmpro.com/news/2014/04/fda-approves-zykadia-late-stage-lung-cancer

17 Food and Drug Administration. White paper: FDA and accelerating the development of the new pharmaceutical therapies [Internet]. U.S. Food & Drug Administration; Mar 23, 2015 [cited Jan 27, 2016]. Available from: https://www.fda.gov/AboutFDA/ReportsManualsForms/Reports/ucm439082.htm

18 The personalized medicine report 2017: opportunities, challenges, and the future [Internet]. Personalized Medicine Coalition; [cited Feb 2017]. Available from: http://www.personalizedmedicinecoalition.org/Userfiles/PMC-Corporate/file/The-Personalized-Medicine-Report1.pdf

19 Gaskin DJ, Richard P. The economic costs of pain in the United States. J Pain. Aug 2012;13(8):715–24.

20 Rudd RA, Aleshire N, Zibbell JE, Gladden RM. Increases in drug and opioid overdose deaths – United States, 2000-2014 [Internet]. CDC MMWR; Jan 1, 2016 [cited Dec 2016]. Available from: https://www.cdc.gov/mmwr/preview/mmwrhtml/mm6450a3.htm

21 Tefferi A, Kantarjian H, Rajkumar SV, Baker LH, Abkowitz JL, Adamson JW, et al. In support of a patient-driven initiative and petition to lower the high price of cancer drugs. Mayo Clin Proc. Aug 2015;90(8):996–1000.

22 Local coverage determinations [Internet]. Centers for Medicare and Medicaid Services; Nov 2, 2015 [cited Feb 1, 2016]. Available from: https://www.cms.gov/medicare/coverage/determinationprocess/LCDs.html

23 CPT codes for molecular pathology [Internet]. Mayo Clinic; 2012 [cited Apr 21, 2017]. Available from: https://news.mayomedicallaboratories.com/2012/12/19/new-cpt-codes-for-molecular-pathology/

24 Gottlieb S. Medicare nixes coverage for new cancer tests [Internet] [cited Apr 21, 2017]. Available from: http://www.forbes.com/sites/scottgottlieb/2013/05/22/medicare-nixes-coverage-for-some-new-cancer-tests/

25 Guillama NJ. Back to the future on ICD-10? [Internet]. PWeR Inc.; Jun 5, 2015 [cited Feb 1, 2016]. Available from: http://www.pwer.com/pwer-news/2015/June/back-to-the-future-on-icd-10

26 Qin C, Zhang C, Zhu F, Xu F, Chen SY, Zhang P, et al. Therapeutic target database update 2014: a resource for targeted therapeutics. Nucleic Acids Res. 2014;42:D1118–23.

27 Chirieac LR, Dacic S. Targeted therapies in lung cancer. Surg Pathol Clin. 2011;3(1):71–82.

28 Transitioning to ICD-10 (Updated) [Internet]. Health Affairs – Health Policy Briefs; [cited Apr 21, 2017]. Available from: http://www.healthaffairs.org/healthpolicybriefs/brief.php?brief_id=111

29 Bowman S. Why ICD-10 is worth the trouble. Journal of AHIMA. Mar 2008;79(3):24–9.

30 Doshi S, Ray D, Stein K, Zhang J, Koduru P, Fogt F, et al. Economic analysis of alternative strategies for detection of ALK rearrangements in non small cell lung cancer. Diagnostics. 2016;6(1):4.

31 NICE diagnostics guidance [Internet]. National Institute for Health and Care Excellence; [cited Feb 1, 2016]. Available from: https://www.nice.org.uk/Media/Default/About/what-we-do/NICE-guidance/NICE-diagnostics-guidance/Diagnostics-assessment-programme-manual.pdf

32 Bücheler M, Brüggenjürgen B, Willich S. Personalised medicine in Europe—enhancing patient access to pharmaceutical drug-diagnostic companion products [Internet]. The European Personalised Medicine Association; Nov 2014 [cited Feb 1, 2016]. Available from: http://www.epemed.org/online/www/content2/104/107/910/pagecontent2/4339/791/ENG/EpemedWhitePaperNOV14.pdf

8

Precision Medicine around the World

Latin America

Felipe is walking home after watching a nail-biting basketball game at his friend's house in the Liberdade neighborhood of Salvador, Brazil. It was the semifinals of the Rio Olympic basketball tournament, and the US team edged out a win against Spain. His favorite player had scored 14 points, the second highest on the US team! Felipe is feeling good. Not just about the game but about things overall. He is happy, even proud, that the games have gone smoothly so far. It has helped him take his mind off of some of the everyday issues that are not as positive in his country, such as the spread of Zika and the continuous political unrest. Sure, Felipe knows there are problems. But he is happy that the Olympics has shown a positive side of Brazil.

When he gets home, Felipe immediately knows that something is wrong. His 60-year-old father, Eduardo, is slurring his speech and experiencing loss of vision in his left eye. Felipe runs to get Ms Costa, their community health agent who lives a few blocks away [1, 2]. Ms Costa typically follows heart failure patients, not stroke patients, in a new pilot program, but Felipe knew she would know what to do. Ms Costa quickly brings Felipe and Eduardo to the local health center. The doctor at the health center gives Eduardo some aspirin and tries to schedule an appointment for an emergency CT scan. There are not very many CT systems in the city of Salvador, and the nearest one takes 2 hours to get to by ambulance. Once there, they wait an additional 3 hours before the machine becomes available. When the scan is finally done, it confirms that Eduardo experienced a stroke. He arrived too late to be given a tissue plasminogen activator (TPA), which helps dissolve clots and increases chances of recovery but has to be given within 3 hours of stroke. Eduardo therefore fell into the "watch and wait" category and stays in the hospital for 4 days.

When Eduardo returns home, Ms Costa comes by to check on him. She goes over the medication that he must take and tells Felipe the schedule so he can remind his father to take his medicine. She also talks to Eduardo about quitting smoking and eating better to avoid strokes in the future. Lastly, she helps him set up home visits from a physical therapist who is part of her team of doctors and

Personalizing Precision Medicine: A Global Voyage from Vision to Reality, First Edition.
Kristin Ciriello Pothier.
© 2017 John Wiley & Sons, Inc. Published 2017 by John Wiley & Sons, Inc.

nurses. Over the next few weeks, she stops by frequently to check on Eduardo and make sure he is recovering.

Felipe's buoyant mood from the Olympics is gone. He is angry that it took so long to get the CT scan for his father. The doctor had told him that results are typically better with TPA rather than "watch and wait." Felipe can't stop thinking that his father would be better if they had gotten the CT scan sooner. Or maybe, if he'd had regular checkups and had been tested already for some of the emerging biomarkers for stroke, including increased C-reactive protein (CRP) or lipoprotein-associated phospholipase A2 (Lp-PLA 2 or the PLAC test) [3], they would have been able to diagnose him faster. His father's father had died of a suspected stroke. Felipe can't help but think this could have been avoided.

Felipe's mother tries to soothe him. She tells him that things have gotten much better. When Felipe's grandfather had his stroke, there was no Ms Costa and no medical help. There were no advanced machines like CT systems back then, and they had to pay for everything themselves, instead of being covered by the SUS, Brazil's universal health plan. Felipe understands his mother's point of view, but he is not consoled.

Latin America is a region that encompasses Central and South America, made up of over 20 countries with over 620 million people. In fact, two of the most populous cities in the world, Mexico City (>21 million people) [4] and São Paulo, Brazil (>26 million people) [5], are in Latin America (Figure 8.1). It stretches from the northern border of Mexico to the southern tip of South America, including the Caribbean, making it one of the most beautiful places in the world for taking in nature, dance, art, and good eating. However, according to UNICEF, Latin America and the Caribbean region has the highest combined income inequality in the world with a measured net Gini coefficient of 48.3, an unweighted average that is considerably higher than the world's Gini coefficient average of 39.7. Gini is the statistical measurement used to measure income distribution across entire nations and their populations and their income inequality [6]. This statistic filters down into all aspects of the region's quality of life, including access to medical care.

As Felipe and his mother's perspectives on Eduardo's stroke illustrate, many people in Latin American are now living longer, healthier lives than their parents or grandparents. In fact, the average person born today can expect to live until the age of 75 years, 6 years longer than someone born 20 years ago. They are also half as likely to die before they turn one. This means that more parents like Eduardo and his wife are celebrating their children's birthdays.

Many of these improvements are due to stronger economies and government policies that have expanded healthcare coverage to the poorest people in the region. Plan Nacer in Argentina, started in 2004, provides basic insurance and secure access to healthcare services for over one million previously uninsured pregnant women and children [7]. Mexico's Seguro Popular now covers over 50 million people, primarily those who are not permanent

Latin America Facts

Definition: 20 countries in Central and South America

Population: 626,741,000 (2015 est.)

Cancer Incidence: 1,096,100(2012 estimate, IARC Global Can

Map:

Figure 8.1 Latin America's demographic facts.

workers, and provides over 200 treatments free of charge [7]. Despite these advances over the past few decades, significant challenges remain in Latin America in terms of giving all people access to quality healthcare and in particular precision medicine.

While healthcare plans have been expanded in Latin America to reach many previously uninsured people, about 54% of the population, or 320 million, still do not have coverage [8]. In other words, a group of people roughly equal to the entire population of the United States either can't go to the doctor or must pay for services out of pocket. Furthermore, the rich, urban neighborhoods get much better healthcare access and treatment than poor, rural areas. Doctors, hospitals, and supplies are concentrated in big cities. Sick people who live outside of the major urban centers must travel at great length and expense for treatment. Imagine waking up at five in the morning to take the bus for 3 hours to a doctor's appointment and then having to come back again the next week for a lab test!

Having enough money to pay for healthcare systems is another challenge. Financial and political challenges have impacted many governments in Latin America during the past few years. Slower economic growth caused by less demand for commodities, combined with a strengthening American economy, has led to weaker currencies compared with the dollar [9]. As a result, it is now more expensive and even unaffordable for many Latin American countries to buy medical supplies and devices paid for in dollars. As standards of living improved for many in recent decades, people started moving less, eating more, and picking up smoking and drinking. These factors, combined with people living longer, mean that more people are now suffering from noncommunicable, chronic diseases such as hypertension, diabetes, and cancer. In 2010, the top three noninjury-related causes of "loss of healthy life" in Latin America were noncommunicable diseases, each up at least 35% since 1990 [10]. The health systems of the region are not ready to deal with this shift. In 2014, noncommunicable chronic diseases accounted for approximately 74 and 81% of total deaths in Brazil and Argentina, respectively [11]. More people are dying of cancer in Latin American countries compared with those in other regions of the world. The problem is only getting worse. An estimated 1.6 million cases of cancer will be diagnosed in 2030, and more than one million cancer deaths are projected to occur annually [12]. To compare the predicament in Latin America versus in other parts of the world, 60% in Latin America of the diagnosed cancer cases result in deaths versus 35% in the United States and 43% in Europe [13].

In light of these challenges, especially the shift toward more chronic disease, the idea of precision medicine is especially important in the context of Latin America. Effectively dealing with chronic diseases usually begins with preventative care and medical tests to understand what is going on with the patients. Preventative care, imaging, and diagnostic testing are some of the key components of precision medicine presented in previous chapters. Currently many Latin American health systems are focused on reactive hospital care, with limited focus on preventative primary care. There are also not enough medical imaging devices or laboratories within most Latin American countries to offer imaging and diagnostics services to the majority of the population. Lastly, only the rich and privately insured can afford or have access to the advanced, targeted treatments based on factors such as genetic markers for certain cancers.

However, in recent years, there have been a number of wins in the field of precision medicine in Latin America. Public–private partnerships have made advanced imaging equipment and IT health solutions available for hospitals in smaller, less affluent cities. New technology such as health-related smart-phone applications and portable X-ray machines are becoming more widely available and adopted. Many of the wins are concentrated in Brazil, Mexico, and Argentina, three countries that receive 92% of all research and develop-ment (R&D) investment in the region [14]. As these countries set the stage for the advancement of precision medicine in the region, it is important to understand the programs, policies, and context of each country in further detail, so let's dig in.

Brazil

Brazil is the most populous country in Latin America with 206 million people [15]. That's roughly two-thirds of the population of the United States. It also has the region's largest economy. Agriculture, mining, manufacturing, and services are all major industries in the country. Brazil experienced strong eco-nomic growth in the 2000s and early 2010s due to China's high demand of their exports. Millions of Brazilians were lifted out of poverty as the middle class grew rapidly. The country was constantly in the international spotlight as it hosted the 2014 World Cup and the 2016 Olympic Games. The good times did not last, however. Now Brazil is in the middle of a deep recession. GDP contracted by 3.8% in 2015 [16]. Political upheaval dominated international news. Millions of activists took to the streets in the spring of 2015 to protest the economic conditions and allegations of corruption involving the Brazilian petroleum company Petrobras and the administration of President Dilma Rousseff. In March of 2016, after spending a week visiting hospitals and labo-ratories in São Paulo, my colleague and I barely got to the airport to fly back to the United States because we had to circumvent the city to avoid protest crowds. Rousseff was eventually impeached and removed from office in August 2016.

Similar to the economy, Brazil's healthcare system has also seen a mix of successes and challenges. In 1990, the country implemented one of the largest universal healthcare systems in the world, known as the Unified Health

System (SUS). Since then, infant mortality has fallen by more than 50%, and life expectancy rose from 66 years in 1990 to 74 years today [17]. Communicable diseases such as tuberculosis are no longer an everyday worry due to earlier detection and improved access to treatment [11]. Yet gaps remain between the rich and the poor. Infant mortality is twice as high in the less affluent northern regions than in the south. The 25% of the population that can afford private health insurance has access to double the number of doctors compared with the rest of the people that rely on the SUS [18]. According to physicians I spoke with, Brazilians that use the SUS for coverage have to deal with shortages of hospital beds and long wait times for basic diagnosis and treatment.

And more oncologists in general are still needed; 1.2 oncologists exist for every 100,000 people. While on the higher end of some of the countries profiled in this book, it still does not have the access strength of the United States or Europe, and Brazil's oncologists are clustered in major cities in specific institutions not accessible to all, making its need to build its oncology programs and oncologist numbers across the country important [19].

Within this context, Brazil has made a number of advancements that set the stage for future developments in precision medicine. To address equity issues and emphasize preventative medicine for all citizens, an innovative program known as Family Health Strategy (FHS) was launched in the 1990s. Rather than forcing low or no income people to travel great distances to see a doctor, this program brings the doctor to the people. FHS puts together interdisciplinary teams that include doctors and nurses but depends on community health agents that live in the communities they serve. Agents visit each household within their area at least once a month. They make sure families are keeping their appointments and taking their medication and look out for risk factors of chronic diseases like diabetes and hypertension. The overall team is responsible for providing preventative health services such as breastfeeding support, prenatal care, immunizations, and disease screening. Each team covers an area of up to 1000 households. The program has been gradually expanded since it started. By 2014 there were 39,000 FHS teams, including 265,000 community health agents that provide services for over 120 million people. That is over 60% of Brazil's population. FHS teams target the poorer segments of municipalities and represent a significant advancement to improving access to preventative care [20].

With programs like FHS improving the infrastructure for primary and preventative care, it will become easier in the future for the broader population to have access to diagnostics and imaging services needed to administer precision medicine. Along this front, Brazil has also made important strides through public–private partnerships. Former President Rousseff in 2015 allowed foreign companies to invest in private hospitals [11]. This led to a number of initiatives, including a partnership between the International

Finance Corporation (IFC) and a consortium of Brazilian diagnostic imaging companies. This partnership invested over US$40 million to improve access to advanced diagnostic tests for underserved areas in the Bahia region of Brazil. A new diagnostics center at the State Center of Oncology in Salvador was built, and 45 pieces of new equipment, including CT scan and MRI machines, were placed in 12 regional hospitals [21]. The diagnostics center provided 180,000 diagnostic tests to low income patients covered by the SUS in its first year. It is expected to eventually serve up to six million patients in Bahia [22].

With steps taken toward expansion of preventative care, diagnostics, and imaging, the final hurdle for precision medicine is access to targeted treatments specific to the patient and the disease. Here, Brazil has shown initiative and support for the idea of precision medicine. In 2015, five research, innovation, and dissemination centers (RIDCs) agreed to collaborate in the Brazilian Initiative on Precision Medicine (BIPMed) [23]. The groups plan to develop a computer platform that brings together genetic data collected by the RIDCs. The platform can then be accessed by researchers in Brazil and internationally. The ultimate goal of the platform would be to advance R&D of drugs that are targeted for specific individuals based on their DNA.

BIPMed was not Brazil's first major foray into the field of genetics and precision medicine. In fact, the country has had a long history of genetic research starting in the 1950s [24]. Since then, Brazil has developed a network of public and privately funded genetic research centers, including the Hugh Genome Center and Stem Cell at the University of São Paulo (USP) and the Molecular Oncology Center at the Hospital Sirio-Libanes (HSL).

The USP has blossomed over the last years into a major center with its genomics center and its extensive pathology programs. USP has 4000 beds including the largest cancer hospital in LatAm, Institute of Cancer of São Paulo (ICESP), with 500 beds. Dr Venancio Alves, professor and chairman of the Department of Pathology, and his team at USP devote their time to teaching and training of many of the pathologists who then make their careers there at USP or in the private commercial labs like Diagnostika (Pardini) or the other large cancer-focused hospitals such as AC Camargo. They also have one of the largest necropsy programs in South America, as they receive all of the bodies from Hospital de Clinicas (one of the largest public hospitals in Brazil) in addition to their own. (Necropsy is another word for autopsy, and this team has quite a program, with the number of bodies coming in, the influx of tools at their disposal, and an under-city set of pathways allowing easy access from wherever the patient dies to where they need to end up for the necropsy that our cab driver first told us about and is apparently quite famous!)

Researchers from HSL are developing new tests that can detect cancer at an earlier stage based on genetic analysis of tumor characteristics of individual patients [25]. Its laboratories are pristine, and its cancer treatment center on one of the top floors caters to its patients by providing not only premiere precision medicine but also its ambience; there are vaulted ceilings, light-bathed private rooms with plenty of seating and comfort for family and friends, and a feel that is more like a spa than a hospital. Other hospitals with top-tier cancer treatment, like Hospital Israelita Albert Einstein, also in São Paulo, provide everything from the top combination medicines to a piano player in the lobby to lighten the atmosphere. Furthermore, Brazil is one of the few countries in Latin America to have a national tumor and DNA bank. The bank stores patient data and tissue samples that have been donated by cancer patients [26]. Scientists can use the samples to support research into identifying genetic markers for cancer that can be used for new drug development. BIPMed and these initiatives are evidence of acceptance of the idea of precision medicine among Brazil's government and medical community.

With improvements in access to preventative care, public–private partnerships supporting availability of diagnostics and imaging equipment, and investment in the idea of targeted cancer treatments, Brazil has the right pieces in place to support precision medicine. Healthcare expenditures have grown steadily, increasing from 6.5% of GDP to 8.3% between 1995 and 2014 [17]. This means increased investment in healthcare resources and potential greater access to higher cost advanced treatments. And its leadership believes in this forward progress not only in basic patient care but beyond. "In limited resource areas, where the idea of 'precision medicine' is still not completely known... it can be the most appropriate solution for a better and more sustainable future of the health sector...," says Carlos Gouvea, president director of Aliança Brasileira da Indústria Inovadora em Saúde (ABIIS) and executive secretary of The Brazilian Chamber for *In Vitro* Diagnostics (CBDL). ABIIS, which consists of the organizations ABIMED, ABRAIDI, ADVAMED, and CBDL, help bring together the medical technology industry in Latin America and worldwide (their definition of medical technology consists of medical devices, medical equipment, diagnostics, and e-health). "If we consider the current revolution generated by the Digital Health / IoT - Internet of Things, we will be able to limit the use of unnecessary and ineffective treatments, allowing customization of care based on prevention. The optimization of existing resources will be leveraged to the maximum level with such tools." ABIIS's 2015 report, entitled Health 4.0: Proposals to boost the cycle of innovation in Medical Devices (MedTechs) in Brazil, highlights that most investments in healthcare are focused on patient care and information infrastructure systems "tend to play a secondary role" [27]. The publication addresses public and private healthcare sectors, development agencies, government, NGOs, and professional organizations to unite different stakeholder views and promote access to new medical diagnostics and devices in Brazil.

Did You Know?

The medical technology industry in Brazil consists of 14,482 companies. Of these, 4,032 are manufacturers and 10,450 are engaged in marketing and distribution of medical technology products. Exports in 2013 totaled US$825 million and represented 15% of Brazilian production of health products. Yet total medical technology expenditure in Brazil, the largest country in Latin America, is in the single-digit percentages (between 2 and 7% based on what is included or excluded in the category) as a percent of total healthcare spending despite all of these companies in the country [27]. High taxes on equipment, delays in Brazilian regulatory approvals, and lack of budget for medical equipment are several of many reasons for the dichotomy. Brazil can produce the medical technology it needs but in many areas is unable to access it.

Premiere national clinical diagnostics reference laboratories such as Fleury and DASA are also leading the way in innovative cancer diagnostics highly connected to the most esteemed medical institutions in Brazil. While visiting a DASA facility in São Paulo, the largest clinical reference lab in Latin America, I was led up into an observation room almost like a scene in one of the *Star Wars* movies. The observation room overlooked one large stadium-like room filled with state-of-the-art diagnostic equipment and track systems all running the tests of the day, with a lean and efficient number of staff suited up to flow effortlessly among the machines providing to the minute quality assurance. From immunoassay to clinical chemistry to hematology to molecular, DASA processes 55,000 test results for patients a day and runs approximately 10 million samples per month [28]. Moving from this environment to the elite and innovative campus of Fleury, armed with the minds of laboratorians such as Edgar Rizzatti, director of Technical Operations, a Fleury veteran for 10 years, was a reminder of all that Brazil is offering in medical research with inside talent and outside partnerships as well. For example, in 2014 Fleury partnered with Veracyte, a US diagnostic company, to offer precision medicine diagnostic Afirma to reduce unnecessary surgeries in thyroid cancer [29]; this offering added to their thorough oncology research and clinical programs in precision medicine. More recently, Fleury has partnered with USP and Heart Institute (InCor) to research diagnosis of cardiomyopathies through next-generation sequencing [30], taking precision medicine beyond oncology. These are just two of the large diagnostic labs serving the Brazilian market and committing themselves to precision medicine. Multiple companies committed to precision medicine, such as Roche and Novartis, also have large divisions in Brazil committed to access and dissemination of the best precision medicine to their populations in need.

But perhaps one of the most impactful institutions in this market today belongs to one of the oldest centers and frequently pointed to as the gold standard in cancer care, AC Camargo. Most of the above institutions follow the lead of this 500-bed hospital for clinical cancer care and diagnostic testing, with a mix of public paying, private paying, and self-paying patients. Precision medicine diagnostics are run by Dr Isabela Werneck de Cunha, medical coordinator of molecular pathology. Precision medicine diagnostics innovation is fueled by the leadership of Dr Dirce Maria Carraro, leader of the genomics group. As we met with them, late, of course, after getting lost between buildings, miming our way through security with minimal Portuguese and tunneling through the original garage turned expansion hospital because AC Camargo grew over the years in São Paulo with no actual room to grow, their enthusiasm and commitment to bring the best of precision medicine to AC Camargo patients was palpable. They especially highlighted their partnerships with pharmaceutical companies to help fund diagnostic testing so that the most precise patients can funnel to the right drugs. In Brazil, lab dependence on this mechanism of funding support is essential to fuel innovation. Dr Fernando Soares, the seasoned and wonderfully energetic CEO and director of pathology at AC Camargo, reiterates this point to be thoughtful of the real São Paulo, a city that makes up about 45% of the people in Brazil. "There are a lot of contrasts in São Paulo. We have to remember that 20–40 million people may have access to centers across the city for cancer care, but 160 million people are struggling in daily life, even with all of this innovation around us." He mentions the growing satellite offices for care that are expanding outside the city as a specific initiative to help this population in need. "This necessary expansion to reach more people closer to where they live, especially in a city like São Paulo where it may take 3 hours to go across the city in our traffic, is a welcome movement in Brazil."

Through continued government and private support, buttressed by initiatives like those from the stakeholders mentioned above, precision medicine in Brazil will continue to blossom, for all patients.

Mexico

Mexico is home to 123 million people and represents Latin America's second largest economy [15]. The country has experienced steady economic growth since 2008. It is a large oil exporter and major supplier for the US manufacturing industry. While some have benefited from the growth, the gap remains large between the rich and poor, as more than 50% of the population lives below the poverty line [15].

Lack of equity is also a major issue facing the country's healthcare system. Prior to 2003, almost 50 million people did not have access to health insurance [31].

These were mostly unemployed, self-employed, or non-salaried informal workers. They did not qualify for coverage under the Mexican Social Security Institute (IMSS), which was for private sector, formal, and salaried workers. They also did not qualify for coverage under the Institute of Social Security and Services for Civil Servants (ISSSTE) for government employees and their families. Then, in 2003, the Seguro Popular or Popular Health Insurance (PHI) was created to cover this segment of the population. Since the establishment of the Seguro Popular, some health outcomes in Mexico have shown improvement. Infant mortality dropped by 38% between 2003 and 2013, and the number of people reporting that health expenses were forcing them into poverty was reduced from 3.3 to 0.8% [15].

Yet despite the fact that access to some healthcare coverage has expanded, many challenges remain. Mexico has fewer doctors, hospital beds, and other resources compared with its peer countries. There are 2.2 physicians and 2.7 nurses for every 1000 inhabitants in Mexico, which is lower than the average of 3.2 physicians and 8.7 nurses among countries that are part of the Organization for Economic Cooperation and Development (OECD, 35 countries in North and South America, Europe, and Asia Pacific) [32]. Mexico also has two to three times fewer specialists, such as oncologists and surgeons, compared with its OECD peers [33]. From a facilities' perspective, in 2011 there were 15 hospital beds per 10,000 people. This is only slightly higher than the ratio in Iraq and roughly half compared with the United States [17]. These capacity problems are made worse by the fact that the majority of resources are concentrated in big cities. Of the approximately 270 oncologists in Mexico, 60% work in the urban centers of Mexico City, Monterrey, and Guadalajara [8]. Someone living in the Federal District (Mexico City) has access to approximately three times more hospital beds compared with someone living in a poorer state [11]. At a presentation on innovation in diagnostics in Mexico City in 2015, I was speaking with an audience of diagnostic developers and their customers. Their major complaint wasn't lack of innovation; it was lack of access to get the innovative equipment to areas of Mexico despite their best efforts. One customer from a small medical institution in Mexico City said, "Kristin, you talk on all of this innovation worldwide. But we can't even get basic lab equipment to some areas of our country. We have to start at a level more basic than you are talking about." This very true (and very humbling) comment underscored the issues Mexico faces.

Given the disparity in doctors, facilities, and equipment between urban and rural areas (and sometimes even within urban), a number of initiatives have focused on expanding access and improving preventative care measures for the underserved. CASALUD, a mixture of the words "house" (casa) and "health" (salud), was established in 2008 as a program focused on detecting and preventing chronic diseases. Similar to the FHS in Brazil, the program shifts healthcare delivery from primary care centers to homes and communities. CASALUD does this through the use of mobile health tools. A community health worker can use MIDO backpack, a tablet-based program, to help assess chronic diseases as she visits people in communities. An individual with diabetes can use MIDO Mi Diabetes, a mobile app, to self-monitor his condition. He enters in his measurements like weight and blood glucose and instantly receives feedback. After initial pilot programs showed successful outcomes in patient self-management and chronic disease management, the program was expanded to more than 20 states and now serves 1.3 million lives each year. CASALUD is run through the Seguro Popular program and therefore covers the segment of the population that is most in need of improved preventative care [33].

Also on the issue of preventative care, between the years 2001 and 2012, which covers the period before and after Seguro Popular was implemented, a greater proportion of older Mexicans were receiving preventative screenings and vaccinations [32]. But, more work is still to be done. Although most of the study subjects were screened for diabetes and cardiovascular disease, many were not screened for many types of cancer and other advanced diseases [32].

Part of the reason is lack of necessary equipment. As of 2011, Mexico had 1.4 MRI machines and 3.6 CT systems per million inhabitants, much lower than many of its OECD peers [34]. In Guerrero, one of the poorest states in Mexico, there were zero CT systems and only one mammography unit available at general hospitals and health care centers [34]. The three million people living here have limited options for breast cancer screening and other diagnostics.

In addition to insufficient healthcare, human resources, and facilities, the segmentation of Mexico's health insurance systems also presents a challenge. Coverage and quality of treatment are by no means equal across the subsystems (IMSS, ISSSTE, Seguro Popular). Certain diseases that are more complicated or expensive to treat, such as lung cancer, are not covered under Seguro Popular [35]. Furthermore, the existence of the segments themselves is an issue. Each system has its own network of doctors, clinics, hospitals, pharmacies, and treatment centers. Patients cannot easily transfer their records between systems [35]. If a formally employed worker loses his job and starts working part time, he will have to see doctors within a new network. His new doctors will not have access to his records from his old network. They won't know what medications he is on or what conditions he's had in the past. His new insurance might not even cover the medications he needs. This makes it challenging to provide comprehensive and continuous care for a person as he rotates between formal and informal labor markets.

In recent years, attempts have been made to reduce the distance between the healthcare program segments. Proposals have been presented to the government to ease the portability, or transfer of medical records, between systems. Seguro Popular plans to expand its oncology coverage to include colorectal cancer in addition to breast cancer [36]. Making the coverage more similar between the healthcare systems will also make it easier for people to transition between them. Yet in order for this expansion to be successful, funding must be improved. In a survey of 128 Mexican oncologists, almost two-thirds of them said that lack of funding from the public health insurance programs prohibited them at some point from prescribing a type of chemotherapy that was known to be effective for breast cancer patients with a certain genetic marker [37].

As in Brazil, a number of advancements have been made in Mexico that lay the groundwork for the future expansion of precision medicine. Programs like CASALUD are expanding healthcare access and shifting the focus of healthcare to preventative care for many living in poorer regions of the country. Seguro Popular is slowly expanding to cover more complex diseases and cancers. With continued investment in healthcare resources such as diagnostic equipment, integration of the separate healthcare programs, and increased funding for the healthcare systems to cover more advanced, targeted treatments, Mexico shows promise for precision medicine in the future.

Argentina

Argentina is the fourth most populous country and the third largest economy in Latin America. The country has abundant natural resources and energy with large agriculture and manufacturing industries. The economy has been a rollercoaster over the past decade, with GDP growing by 8.4% in 2011 then contracting −2.6% in 2014 [17]. Inflation has been consistently high as a result of government macroeconomic policies.

Argentina's healthcare system is organized into three sectors. Formally employed workers are covered by plans managed by workers unions, or Obras Sociales, which are funded through payroll contributions. More affluent members of the population can purchase private health insurance. The remainder, and often those who are unemployed or work in the informal sector, depends on the public health system for care. Argentina's healthcare system faces many of the same challenges as discussed for Mexico, including lack of integration between the three sectors and inequalities in access and quality of care.

One of the major advancements in Argentina related to precision medicine was in the improvement of access to primary and preventative care through Plan Nacer. The plan was launched in 2004, when the economy was in crisis and many families were struggling. The plan provided public health insurance to uninsured pregnant women and children under the age of 6. Pregnant women were offered services for prenatal care, delivery, and postnatal care. Children were given neonatal care, immunizations, nutrition support, and disease treatment until the age of 6 years. To encourage local governments to improve access and quality to this target population, the plan used an innovative pay for performance scheme. Provincial or local governments were paid fixed amounts based on the number of people enrolled in the plan as well as for achieving health indicator targets. Studies now show that Plan Nacer was effective at improving birth outcomes, reducing death of newborns in hospitals by 74% for those enrolled [38].

In 2012, the government launched Plan Sumar, which extended Plan Nacer to cover children and adolescents under the age of 19 years and uninsured women between 20 and 64 years. Services offered were also expanded to meet the needs of this different demographic, but similar pay for performance mechanisms were in place. Plan Sumar is expected to cover 5.7 million children and 3.8 million women [39].

A second advancement related to precision medicine in Argentina was the establishment of the National Cancer Institute (NCI) in 2010. The creation of this institute put cancer at the top of the government's agenda. The role of the NCI is to reduce incidence and mortality from cancer and to develop public policies related to prevention, diagnosis, and treatment of the disease [40].

Similar to Mexico, Argentina has in recent years made efforts to improve access to preventative and primary care and laid the groundwork for future emphasis on targeted cancer treatments through the establishment of NCI. In order for precision medicine to grow in Argentina, the government must invest more into healthcare expenditure, work to decrease the fragmentation between healthcare programs, and focus on improving the country's diagnostic and imaging infrastructure.

South America: Beyond

Evidence of opportunities for precision medicine can also be seen outside of Brazil, Mexico, and Argentina. These have largely been driven by public–private partnerships and efforts by large, multinational healthcare organizations. In 2015, Pfizer launched the Center of Excellence for Precision Medicine (CEPM) in Chile. Together, the government and the company are investing $21 million to create a regional research and development hub [41]. CEPM will first focus on developing treatments for lung cancer, which is becoming an increasingly large problem for many Latin American countries.

Costa Rica is another Latin American country that is up and coming on the precision medicine front. With a population of only 4.7 million, but the proportion of working age people over 66% of the population, this vibrant country known for ecotourism and zip lining adventures is also an aspiring hub for medical innovation. A new law enacted in 2014 lifted a 4-year ban on human medical research, and the country is looking to regain the trust of investors and encourage growth and clinical trial activity for pharmaceuticals in the country. It is already a major manufacturer for medical devices, as the second largest exporter of medical devices in Latin America after Mexico. The country has over 60 medical technology companies, including Philips, Medtronic, Terumo, and Steris, among others, producing class I to class III medical devices, with exports growing over 190% in the last decade to become the number 1 export product (23% of Costa Rica's exports) in 2015 [42].

Many of those company's manufacturing sites live in the Coyol Free Zone (CFZ), a one-stop-shop bio-park set in Alajuela. When visiting CFZ, I was struck by the organized maze of device manufacturers all set in such a lush, green, open setting, different from bio-parks in other areas of the world and in stark contrast to one I was visiting in France a month before. When going from one company to another, the vision of each specific company was highlighted but equally important was the attention to the local employees and their needs. CFZ offers preconstruction, execution, and post-construction services all to meet top global regulatory requirements as company manufacturing sites are being built. In addition, they offer "Gente Coyol" services that provide all 8000 employees working in various companies in the park wellness, transportation, and other employment services, all specialized to their specific population needs [43]. Costa Rica, in addition to producing precise medical diagnostics and devices for nations worldwide, is also paying attention to providing precision wellness for its workers developing those devices.

As you have seen, the growth of precision medicine in Latin America faces numerous challenges. Lack of access to preventative care, advanced diagnostics, and high cost medications for a significant portion of the population is a major hurdle to overcome. Yet some countries are taking steps forward, and all are excited to continue. Expanding preventative care through innovative community-based programs, leveraging public and private partnerships to access more advanced technologies, and investing in research for targeted treatments of advanced diseases are among the trends that can lead to future growth of precision medicine in the region.

As Felipe despairs over the health of his father, there is hope that he and his children can benefit from future advancements in the field of precision medicine. Progress in this area can build on recent developments in Latin America to drive prosperity and healthcare equality into the future.

References

1 da Silva AF, Cavalcanti ACD, Malta M, Arruda CS, Gandin T, da Fé A, et al. Treatment adherence in heart failure patients followed up by nurses in two specialized clinics [Internet]. Rev Lat Am Enfermagem. Sep–Oct 2015; [cited Nov 14, 2016];23(5):888–94. Available from: https://www.ncbi.nlm.nih.gov/pmc/articles/PMC4660411/

2 Mantovani VM, Ruschel KB, de Souza EN, Mussi C, Rabelo-Silva ER. Treatment adherence in patients with heart failure receiving nurse-assisted home visits [Internet]. Acta Paul Enferm. Jan/Feb 2015 [cited Nov 14, 2016];28(1). Available from: http://www.scielo.br/scielo.php?pid=S0103-21002015000100041&script=sci_arttext&tlng=en

3 Stroke biomarkers risk [Internet]. Stroke Biomarkes; [cited Nov 14, 2016]. Available from: http://stroke-biomarkers.com/biomarker_list. php?filter_by=endpoint&detail=Risk

4 Mexico City population [Internet]. World Population Review; [cited Nov 14, 2016]. Available from: http://worldpopulationreview.com/world-cities/ mexico-city-population/

5 Estimativ as das populações residentes, em 1° de julho de 2009, segundo os municípios [Internet]; [cited Nov 14, 2016]. Available from: http://www. ibge.gov.br/home/estatistica/populacao/estimativa2009/POP2009_DOU.pdf

6 Ortiz I, Cummins M. Global inequality: beyond the bottom billion [Internet]. UNICEF; April 2011 [cited Nov 14, 2016]. Available from: https://www.unicef. org/socialpolicy/files/Global_Inequality.pdf

7 The World Bank. Universal healthcare on the rise in Latin America [Internet]. Feb 14, 2013 [cited Nov 14, 2016]. Available from: http://www.worldbank.org/ en/news/feature/2013/02/14/universal-healthcare-latin-america

8 Gross PE, Lee BL, Badovinac-Crnjevic T, Strasser-Weippl K, Chavarri-Guerra Y, St Louis J, et al. Planning cancer control in Latin America and the Caribbean. Lancet Oncol. Apr 2013;14(5):391–436.

9 Sambo P, Orr E, Godoy D. Brazilian real drops to record low against U.S. dollar [Internet]. Bloomberg; Sept 22, 2015 [cited Nov 14, 2016]. Available from: https://www.bloomberg.com/news/articles/2015-09-22/ brazil-s-currency-tumbles-to-record-on-pessimism-over-budget

10 Institute for Health Metrics and Evaluation. The Global Burden of Disease: Generating Evidence, Guiding Policy. Seattle, WA: IHME; 2013 [cited Nov 14, 2016]. Available from: http://www.healthdata.org/sites/default/files/files/ policy_report/2013/GBD_GeneratingEvidence/IHME_GBD_ GeneratingEvidence_FullReport.pdf

11 Global Health Intelligence. Opportunities in Latin America's healthcare sector 2016 [Internet]. Global Health Intelligence; Jan 15, 2016 [cited Nov 14, 2016]. Available from: http://globalhealthintelligence.com/ghi-analysis/ opportunities-in-latin-americas-healthcare-sector-2016/

12 Bray F, Piñeros M. Cancer patterns, trends and projections in Latin America and the Caribbean: a global context. Salud Publica Mex. Apr 2016;58(2):104–17.

13 TJCC, Abrale, Department of Education and Research, Lobo TC. The borders between countries are decreasing for the benefit of health [Internet]. Observatorio de Oncologia; Dec 16, 2016 [cited Jan 20, 2017]. Available from: http://observatoriodeoncologia.com.br/ las-fronteras-entre-los-paises-estan-disminuyendo-en-beneficio-de-la-salud/

14 Deloitte Global. Latin America economic outlook [Internet]. Deloitte; Jul 2015 [cited Nov 14, 2016]. Available from: https://www2.deloitte.com/global/en/ pages/about-deloitte/articles/latam-economic-outlook-report.html

15 The World factbook [Internet]. Central Intelligence Agency; [cited Oct 27, 2016]. Available from: https://www.cia.gov/library/publications/the-world-factbook/

16 The World Bank. Brazil Overview [Internet]. The World Bank; [cited Oct 27, 2016]. Available from: http://www.worldbank.org/en/country/brazil/overview

17 The World Bank. DataBank [Internet]. The World Bank; [cited Oct 27, 2016]. Available from: http://databank.worldbank.org/data/home.aspx

18 Kaiman J, Smith D, Anand A, Watts J, Kingsley P, Hooper J, et al. How sick are the world's healthcare systems? [Internet]. The Guardian; Oct 29, 2014 [cited Nov 14, 2016]. Available from: https://www.theguardian.com/society/2014/oct/29/how-sick-are-worlds-healthcare-systems-nhs-china-india-us-germany

19 Goss P. The lancet oncology: commission shows good progress in cancer care in Latin America. EurekAlert [Internet]. AAAS; Oct 28, 2015[cited Apr 13, 2017]. Available from: https://www.eurekalert.org/pub_releases/2015-10/tl-tlo102715.php

20 Macinko J, Harris MJ. Brazil's family health strategy—delivering community-based primary care in a universal health system. N Engl J Med. Jun 4, 2015;372 (23):2177–81.

21 IFC Public-Private Partnerships. Public-Private Partnership Stories [Internet]. Washington, DC: International Finance Corporation; [cited Oct 27, 2016]. Available from: https://www.ifc.org/wps/wcm/connect/75bffe804a9a408d9b8 0df9c54e94b00/PPP+Stories_BA+Health+II_Final+2015.pdf?MOD=AJPERES

22 IFC Public-Private Partnerships. Filling a Critical Health-Care Gap in Brazil [Internet]. Washington, DC: International Finance Corporation; Aug 2016 [cited Oct 27, 2016]. Available from: http://www.ifc.org/wps/wcm/connect/news_ext_content/ifc_external_corporate_site/news+and+events/news/filling-a-critical-health-care-gap-in-brazil

23 Marques F. Precision medicine [Internet]. Rev Pesqui Fapesp. Nov 2015 [cited Oct 27, 2016]. Available from: http://revistapesquisa.fapesp.br/en/2016/03/24/precision-medicine/

24 Passos-Bueno MR, Bertola D, Horovitz DDG, de Faria Ferraz VE, Brito LA. Genetics and genomics in Brazil: a promising future. Mol Genet Genomic Med. 2014 Jul;2(4):280–91.

25 Zorzetto R, Da Silveira E. Catching cancer in the act [Internet]. Rev Pesqui Fapesp; Nov 2015 [cited Oct 27, 2016]. Available from: http://revistapesquisa.fapesp.br/en/2016/03/28/catching-cancer-in-the-act/

26 Gonçalves AA, Claudio Pitassi C, Assis VM. The case of INCA's National Tumor Bank management system in Brazil. J. Inf. Syst. Technol. Manag.. Dec 2014;11(3):549–68.

27 The Websetorial Consultoria Economica Team. Health 4.0 [Internet]. ABIIS; 2015 [cited Oct 27, 2016]. Available from: http://www.abiis.org.br/abiis-health-4.0.html

28 DASA: the biggest medical diagnostics company in Brazil [Internet]. Elga Veolia; [cited Oct 27, 2016]. Available from: http://www.elgalabwater.com/ dasa-biggest-medical-diagnostics-brazil

29 Veracyte. Veracyte and Fleury announce partnership to make the Afirma® gene expression classifier available to patients in Brazil. PR Newswire; May 2, 2014 [cited Oct 27, 2016]. Available from: http://www.prnewswire.com/news-releases/veracyte-and-fleury-announce-partnership-to-make-the-afirma-gene-expression-classifier-available-to-patients-in-brazil-257648951.html

30 Fleury investor day presentation 2016 [Internet]. GrupoFleury; 2016 [cited Oct 27, 2016]. Available from: http://ir.fleury.com.br/fleury/web/ default_en.asp?idioma=1&conta=44

31 OECD. OECD Reviews of Health Systems: Mexico 2016 [Internet]. Paris: OECD Publishing; 2016 [cited Oct 27, 2016]. Available from: http://dx.doi. org/10.1787/9789264230491-en

32 Salinas JJ. Preventive health screening utilization in older Mexicans before and after healthcare reform [Internet]. Salud Publica Mex. 2015;57(suppl 10):S70–8; [cited Apr 13, 2017]. http://www.scielosp.org/pdf/spm/v57s1/v57s1a11.pdf# page=1&zoom=auto,-274,765

33 Thoumi A, Maday M, Drobnick E. Preventing chronic disease through innovative primary care models [Internet]. Brookings; 2015 [cited Oct 27, 2016]. Available from: https://www.brookings.edu/wp-content/ uploads/2015/04/chp_20150407_mexico_casalud.pdf

34 Azpiroz-Leehan J, Licona FM, and Méndez MC. Imaging facilities for basic medical units: a case in the State of Guerrero, Mexico. J Digit Imaging. Oct 2011;24(5):857–63.

35 ManattJones Global Strategies. Mexican healthcare system challenges and opportunities [Internet]. ManattJones Global Strategies, LLC; Jan 2015[cited Oct 27, 2016]. Available from: https://www.wilsoncenter.org/sites/default/ files/mexican_healthcare_system_challenges_and_opportunities.pdf#page=1& zoom=auto,-265,792

36 Taylor L. Brazil, Mexico access to expensive cancer drugs 'improving' [Internet]. Pharmatimes.com; Apr 12, 2012 [cited Oct 27, 2016]. Available from: http://www.pharmatimes.com/news/ brazil,_mexico_access_to_expensive_cancer_drugs_improving_977505

37 Chavarri-Guerra Y, St Louis J, Liedke PE, Symecko H, Villarreal-Garza C, Mohar A, et al. Access to care issues adversely affect breast cancer patients in Mexico: oncologists' perspective. BMC Cancer. Sep 9, 2014;14:658.

38 The World Bank. Argentina's Plan Nacer Delivering Results for Mothers and Their Children [Internet]. The World Bank; Sep 18, 2013 [cited Oct 27, 2016]. Available from: http://www.worldbank.org/en/topic/health/brief/ argentinas-plan-nacer-delivering-results-for-mothers-and-their-children

39 The World Bank. Argentina: Plan Nacer Improves Birth Outcomes and Decreases Neonatal Mortality among Beneficiaries [Internet]. The World

Bank; Sep 18, 2013 [cited Oct 27, 2016]. Available from: http://www.
worldbank.org/en/news/press-release/2013/09/18/argentina-plan-nacer-
birth-outcomes-decreases-neonatal-mortality-beneficiaries

40 Huñis AP. A current view of oncology in Argentina. Ecancermedicalscience.
2016;10:622.

41 Pfizer's Center of excellence in precision medicine in Chile [Internet].
Innovation Insights; [cited Oct 27, 2016]. Available from: http://www.
innovationinsights.ch/pfizers-center-excellence-precision-medicine-chile

42 Life Sciences Costa Rica Life Sciences Website. Essential Costa Rica and
CINDE; 2016 [cited Oct 27, 2016]. Available from: http://www.cinde.org/en/
news/press-release/
life-sciences-forum-2016-costa-rica-will-host-a-world-class-event-for-the-
life-sciences-industry

43 Coyol Free Zone. [cited Oct 27, 2016]. Available from: http://www.coyolfz.
com/index.php/investors

9

Patients as the Poorest Princesses

Supportive Care in Precision Medicine

> You have no idea what a $300 check can mean to someone so they can pay their family's food bill while in a clinical trial.
> —*Patricia Goldsmith, CEO, CancerCare*

I spend most of my time focusing on the precision medicine technology and products on the market or in development that are helping save patients' lives. My teams and I work to help companies determine the best way to get those products to patients. In some countries, this is an almost insurmountable task; in others, there have been tremendous strides. But what we spend less time on overall is what happens around the patient even if they are able to access the products we are trying so hard to get to them. To a patient and the patient's caregivers, of equal importance to these technologies is all of the other support, or "supportive care," needed when a patient is receiving treatment and beyond. When I remember my friend Heather during her treatment cycles, what she spoke of first was not the therapy infusing into her body, nor the surgery that cut the cancer out of her. The first thing was her supportive care. It was her husband, her Soul Cycle classes, her best friends, and her caretakers that helped her through the day. It was getting critical direction by reading informational websites, figuring out what she could eat or would have the strength to do on a daily basis, connecting with support groups, and paying attention to her insurance and how to pay her bills during her care. Supportive care is all of that. When described as a program, Supportive Care is an effort to meet patients' physical, informational, emotional, psychological, social, spiritual, and practical needs during the pre-diagnostic, diagnostic, treatment, and follow-up phases of care [1]. Each patient is different in the intensity of their supportive care needs. In essence, supportive care needs to be as precise as their precision medicine.

Patients dealing with cancer are no different than people who are well. They usually need to work, pay their bills, shuttle around their children or parents or whoever else depends on them, complete household tasks while

Personalizing Precision Medicine: A Global Voyage from Vision to Reality, First Edition.
Kristin Ciriello Pothier.
© 2017 John Wiley & Sons, Inc. Published 2017 by John Wiley & Sons, Inc.

mumbling, and move through all other aspects of daily life. Sure, they may now be treated like "medical royalty," with precision medicine that is personal to their cancer. But when they are sick, being medical royalty does not feel like anything special. It can sometimes feel overwhelming due to the additional logistic, monetary, and other pressures piling on top of their daily lives. All of these other pressures, plus potentially years of adjacent medical needs brought on by treatment or recovery, are included in a patient's supportive care need. As medical care advances and patients survive longer, the scope of supportive care continues to expand, with an increasing unmet need for support services during survivorship. More cancer patients are living longer and, by default, their families, friends, and caregivers are caring for them longer.

Supportive care as a discipline grew out of hospice care over the last two decades. Hospice care typically refers to services provided to the patient (and sometimes her family) directly before death. Over the years, services expanded to provide palliative care, typically focusing on advanced disease leading into death. Finally, as patient numbers expanded and patients were diagnosed earlier, the demand for services in supportive care from diagnosis and early disease into advanced disease burgeoned [1]. Unfortunately, supportive care for precision medicine patients is fragmented across a number of organizations, all with different missions and usually devoted to one type of disease and/or a few types of care. In some parts of the world, like Europe, accredited cancer centers (accredited by the European Society of Medical Oncology (ESMO)) require comprehensive supportive/palliative care services to be provided. But in the United States, the National Cancer Institute (NCI) cancer center accreditation doesn't have specific provisions requiring supportive/palliative care services, leaving it up to the cadre of disparate services across multiple centers [2]. Nonprofit organizations fill gaps that the centers do not provide, but learning of and accessing those centers is not always easy, which also tend to be regionalized.

Supportive care services can be segmented into three major categories: physical and daily living, financial matters, and other supportive care requirements such as psychological needs, sexual needs, and health system/ informational needs. Each of these categories blends into and overlaps with one another, but each is equally important (Figure 9.1).

The physical and daily living category is the one most commonly highlighted with patients facing aggressive treatments including surgery, chemotherapy, and radiation in various combinations. Physical and daily living care includes patient care and family care. For the patient, ensuring access and adherence to all equipment and pharmaceuticals to treat the disease, receiving diet, skin, and body care to treat surgery scars and radiated skin, and additional medication to curb side effects are major examples. For the family or caregiver, getting the patient to and from the infusion center, clinic, or hospital depending on the treatment they need, rearranging their home to accommodate the patient

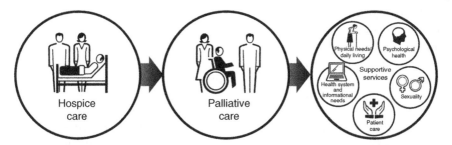

Figure 9.1 The three categories of supportive care services.

moving less or not at all or getting sick frequently during a course of therapy, arranging for visiting nurses or other caregivers during the treatments, and ensuring their children are cared for are major examples.

This category is one for which the majority (over 80%) of care centers in developed countries provide some services or at least guidance, with dedicated cancer centers being the major players. Therefore, any patient living away from a major cancer center needs to carefully arrange their physical and daily living to get by without help, or piece together a collection of helpers to assist them. Numerous studies have pointed out the enormity of this issue, and it hits lower-income and minority populations disproportionately. An example is a study of patients with colorectal cancer in California, with data spanning a 10-year period that shows 88% of minority colorectal cancer patients live more than 5 miles away from a major cancer center and have less successful outcomes as a group [3]. Lack of proximity to a center, in addition to lack of ability to get to the center reliably and inexpensively, contributes to hardships in basic physical care, let alone care tailored specifically to the patient.

In the 1980s, I watched my grandmother and my father contemplate the decision to leave my grandfather (Angelo, or as us children called him, Grandpa "C") with metastatic lung cancer in a rundown long-term care hospital. I was 12 years old, and I was terrified every time we had to go to that sad place, with bored-looking nurses, a dank urine smell, and a deafening quietness except for the moaning of a person down the hall, or the shuffling of feet of the few patients who were able to get out of bed. Even at 12, I knew my grandfather had little time and I wanted him to be at home for his final days, with his jewelry-making tools, his sailing pictures, and the comfort of his family and his things around him. What I didn't understand was that my father and mother were working full time and had my two siblings and me to watch over, my grandmother was 90 pounds and couldn't support my grandfather's weight to move him if he were at home, and their house wasn't set up to accommodate his needs. At the time, there were no programs to help him or them; we were all stuck. He only spent 2 months in that hospital before he died. To this day, not being able to bring him home weighs on my father.

Thankfully, supportive care programs do exist today. The Dana–Farber Cancer Institute in Boston is a strong example of a major medical center with an excellent daily living program to support patients and families. It covers almost every category of supportive care and even some not mentioned. One highlight includes their Leonard P. Zakim Center for Integrative Therapies. This portion of their supportive care offering focuses on innovative and creative complementary therapies into traditional cancer care to improve the quality of life for cancer patients and their families. These offerings include mobile art carts, music and writing therapy, massage and acupuncture, nutritional programs, and movement programs specifically designed for patients with cancer [4]. The programs also incorporate caregivers and other family members in order for them to participate and become more aware of how to modify basic activities for their family members with cancer.

Financial matters are the second major category and one I found most difficult for patients and their families to comfortably talk about when interviewing them for this book. Even at major centers that typically provide the most overall supportive care, a study found about a third of families are reporting food, housing, or energy insecurity 6 months into treatment [5]. One in three families! And even when centers provide guidance or some assistance to help families, there is also a perception issue to take into account. Some patients and their caregivers told me that they felt embarrassed to be "talking about money" throughout their treatment. If they were the patients, they felt they should be talking about their drive to beat the cancer, to get better, to focus on their body, and to survive for their children. If they were the patients' caregivers, they felt pressure to support their spouse or child or friend despite any concerns of how to pay for treatment, or more often, how to pay for everything else while paying for treatment. "It's a terrible thing," said Carla Tardif, Chief Executive Officer at Family Reach foundation, an organization that specifically addresses what it calls the financial toxicity of cancer. "A patient feels like she is panhandling- friends and family bring food but what you really need is the electricity paid and you need a check. Not being able to pay for basic needs can cripple a whole family when they are already struggling both emotionally and physically."

Carla and the Family Reach organization have an intense and needed focus on addressing the financial burden of cancer that is less well covered by the centers that focus on other areas of supportive care. Founded in 1996, Family Reach has been working for 20 years to reduce the financial burden of cancer for families, especially those families with children to care for, either because the patient is a child or a child's parent/caregiver has cancer. In their latest report, they note that cancer hospitalizations cost five times as much as hospitalizations for other complex chronic conditions, with childhood cancer costs reaching almost $32,000 more per hospital stay, and that 14.6% of medical bankruptcy filers indicated the patient was a child [6]. Carla, herself a cancer survivor, spoke with a mix of passion and practicality on family needs while caring for a cancer patient. "The one thing we wish we didn't see as much of is families who waited too long, either because they were too proud to ask for help or were embarrassed to ask. While we can address a variety of financial needs, it is easier for us to address when the need is manageable rather than when the need is catastrophic."

Cofounders Rich Morello and Christopher Wiatrak started Family Reach after they lost Kristine Morello-Wiatrak, Rich's sister and Chris's wife, to cancer. The foundation currently partners with over 165 hospitals and centers to help provide financial information and bridges for families in need. They have developed a financial handbook for families to use proactively to evaluate their financial situation before treatment begins and have financial navigators on staff to support patients or caregivers through their financial journey in cancer [7]. Rich mentions, "We have seen some families that are down to contemplating whether they should call an ambulance or just try to get the patient to the hospital in their own car, they are so worried on expenses, and so ashamed they have to contemplate it. And every family is different in what they are dealing with, so we try to be specific. We try to provide precision financing for precision medicine." Their foundation today supports US families, but it is the type of foundation that needs to be replicated worldwide in order to address this basic supportive care need.

Did You Know?

The Lifeline Grants Program is Family Reach's most utilized assistance program. The program provides monetary grants to cover household bills, including food, childcare, transportation and car payments, mortgage and rent, lodging near treatment sites, and utilities. There are three simple criteria for eligibility: a cancer diagnosis, active treatment within the past year, and financial hardship due to cancer diagnosis. After the criteria are met, financial assistance after processing is within two business days. Payments from the Lifeline Grants Program are sent directly to the designated vendor/business, removing the burden of follow-up from the caretaker's responsibility. Lifeline grants range from $250 to $2000, depending on family needs, and allow families the flexibility to apply the funds where they need it most. Families may apply more than once. They also have other larger grants available for families in grave need, such as those risking homelessness because of the burden of their or their child's cancer costs [6].

The last category of supportive care is a large bucket of many needs, including psychological, sexual, and informational needs for the patient and families during their cancer journey. These are ones that typically fall after basic care of the patient and family and also after payment of care, for obvious reasons.

However, these affect the core of a cancer patient and could affect the overall treatment success and "wellness" of themselves and their families for years after their treatments end.

Psychological and social support overlaps with basic care needs but go beyond that. The psychological ramifications of a cancer diagnosis and journey have a measurable and lasting impact on the patient and his or her family. Many families deal with it privately. Most need the support in the form of either social community support (i.e., support groups, church groups, caregiver groups) or psychological support (i.e., psychological therapy support), or both. Knowing where to go and where the patient and his or her family will be most comfortable and have the most support is crucial.

Another example is sexual needs and the potential to maintain a healthy sex life with one's partner, or even still feel like a normal human, while going through the cancer journey. One patient explained, "You feel badly for being vain, because you are thankful to be alive. But I wanted to feel sexy again. I also wanted my breasts to look normal again and it is taking a long time. With all they had to remove from my right side, they then had to reduce my unaffected left side because the reconstructed side couldn't be built back up. These are the things that they don't tell you, and sometimes you don't want to ask."

A final example is education and the availability of reliable, understandable information about diagnosis, treatment, and supportive care for patients and families. It is not the lack of websites or the volume of data, unless you have no access to the internet, which in many developing and in some developed countries is still a major issue. There are literally thousands of websites covering various aspects of what is in this chapter. But sifting through what is relevant, what is reliable, and what is understandable is the real challenge. Organizations have been established worldwide just to help patients qualify websites on reliable and fact-based information. One of these organizations, which the American Cancer Society highlights on its website [8], is Health On the Net Foundation (HON). HON is a Swiss-based organization that guides patients, physicians, and publishers to reliable medical information online. Websites can bear the HON logo only if they abide by an ethical code of conduct that covers site sponsorship rules, documentation of materials, and authorship. The site and organization also houses reliable medical search engines that filter results by "HONcode certified" websites [9].

If your head is spinning from all of the different organizations involved in supplying supportive care to a patient and his family, you aren't alone. This is a major complaint of patients with cancer. However, there is an organization that has put most of the information from all three areas under one roof and support team. CancerCare is the oldest nonprofit supportive care organization in the United States [10]. Celebrating its 73rd year, it has served over 170,000 individuals and nearly 90% of all counties in the United States. When people come to CancerCare, they may be looking for daily living, psychological support, social and community support, financial support, and education or for something as simple but personal as a hair wig during chemotherapy. The organization is staffed with over forty advanced-degree social workers who can answer calls and assist individuals with any issue they are facing. CancerCare also has a lawyer on staff if patients need representation. CancerCare has one of the largest databases in the United States with a truly holistic look at the patient. In addition to their medical records, CancerCare also houses patients' psychosocial data, 80% of it coming directly reported from the patient.

The organization recently completed its landmark study on patient support on 3000 unique cancer patients that specifically match the demographics of the US population. One of the most impactful findings was that of the surveyed individuals, over 80% had all the information they felt they needed on treatment and the benefits of treatment, but that is where their information stopped. They did not have information on most of their other key elements of supportive care, from side effects, to where to go for caregivers, to what treatment would cost, to whether they could continue working, and so on.

Patricia Goldsmith, CancerCare's vivacious CEO whom I met years ago when we were working on a healthcare informatics project together, launched into our conversation with the strength and practically she brings to the entire organization:

> Look, a patient's average time with a medical oncologist is 12 minutes and they know the oncologist is busy. Patients are reluctant to ask a lot of questions. They are happy the physician has detailed a treatment plan they express confidence in but then they have so many other questions. Thousands of dollars a month even with my insurance? 12 weeks of infusions at a facility not near my house? How can I afford this? How can I take time off without losing my job? Who will care for my kids, get them to and from school, feed them, put them to bed if I am too weak? What can I do if the side effects are so severe I can't take care of my family? How do I explain to my kids about cancer to not scare them but without making false promises about how I will do?

CancerCare is designed to holistically help the patient in answering those questions and providing support. The organization has created online portals so patients can enter an organized group in their community that lets their friends, family, church, or other groups of people close to them sign up and cover dinner, pickups for children, and visiting times. Their factual but understandable articles, also all online, help prepare patients before their appointments and provide additional data to fill in the blanks after their appointments. They provide the entire gamut of supportive care, and anything they do not support fully, and they have the connectivity to get patients to the right sources of care. They are currently based in New York and surrounding areas but are continuing to expand to meet demand.

Dana Farber, Family Reach, and CancerCare all have supportive care offerings that are sorely needed. However, they and the many other organizations worldwide that provide these services to families are not enough. From everyone I talked with, the consensus was the largest challenge in supportive care delivery today is funding. There is never enough funding for these programs and organizations that support these patients wherever they are housed. But they are vitally important to the patient's success on treatment. "Precision medicine or basic care, patients have unbelievably simple unmet needs," says Goldsmith, as she talks about the next steps of CancerCare. They are the needs that, when met, allow the patient and their families to more easily tackle their cancer journey and are as vitally important to their survival as the precision medicine they are taking in.

References

1 Hui D, De La Cruz M, Mori M, Parsons HA, Kwon JH, Torres-Vigil I, et al. Concepts and definitions for "supportive care," "best supportive care," "palliative care," and "hospice care" in the published literature, dictionaries, and textbooks. Support Care Cancer. March 2013;21(3):659–85.
2 Smith TJ, Terrin S, Alesi ER, Abernethy AP, Balboni TA, Basch EM, et al. American Society of Clinical Oncology provisional clinical opinion: the integration of palliative care into standard oncology care. J Clin Oncol. Mar 10, 2012;30(8):880–7.
3 Huang LC, Ma Y, Ngo JV, Rhoads KF. What factors influence minority use of NCI center centers? Cancer. Feb 1, 2014;120(3):399–407.
4 The Leonard P. Zakim Center for Integrative Therapies [Internet]. Dana-Farber Cancer Institute; [cited Nov 23, 2016]. Available from: http://www.dana-farber. org/Adult-Care/Treatment-and-Support/Patient-and-Family-Support/Zakim-Center-for-Integrative-Therapies.aspx
5 Wolfe J. Almost one third of families of children with cancer have unmet basic needs during treatment [Internet]. Dana-Farber Cancer Institute; Sep 23, 2015 [cited Nov 23, 2016]. Available from: http://www.dana-farber.org/Newsroom/ News-Releases/almost-1-third-of-families-of-children-with-cancer-have-unmet-basic-needs.aspx
6 Family Reach survival at all costs: a Family Reach report on the financial burden of cancer [Internet]. Family Reach; Aug 16, 2016 [cited Nov 23, 2016]. Available from: http://familyreach.org/wp-content/uploads/2016/08/16-FamilyReach-Survival-at-all-costs.pdf

7 Family Reach. Family Reach financial handbook [Internet]. Family Reach; [cited Nov 23, 2016]. Available from: http://familyreach.org/financial-edu/

8 American Cancer Society Medical and Editorial Content Team. Cancer information on the internet [Internet]. American Cancer Society; [updated Nov 2, 2016, cited Nov 23, 2016]. Available from: http://www.cancer.org/cancer/cancerbasics/cancer-information-on-the-internet

9 Health on the Net Foundation. [cited Nov 23, 2016]. Available from: www.hon.ch

10 CancerCare. [cited Nov 23, 2016]. Available from: www.cancercare.org

10

Informatics under the Hood

Information in Precision Medicine

> Diagnostics is an information business with a wet lab on the side.
> —*Mara Aspinall, Executive Chairman of GenePeeks and
> Founder of the School for Biomedical Diagnostics at
> Arizona State University*

Thanks to the ever-increasing convergence of data science and biology, we can no longer think about diagnosing a human with any condition, sick or well, without thinking about the ream of data that will result from that single test or set of tests. Since the first human genome was successfully sequenced in the 2000s, the so-called Big Data Revolution has officially begun in the field of life sciences. And this data is not just limited to patient genetic sequencing. DNA provides the cellular instructions for manufacturing proteins, but researchers are also interested in delving down to the individual protein level—and even more granular. It entails deep-dive analyses on ever-growing databases outside of a given patient's biological information, including chemical structure libraries, scientific literature networks, patient data from social media, pharmaceutical company data, and insurance claims data (Figure 10.1). In 2013, an estimated 153 exabytes of data are in the healthcare realm. (An exabyte is a multiple of the unit byte for digital information. One exabyte is one quintillion bytes!) By 2020, there will be an estimated 2314 exabytes of digital information [2]. The task of curating, sanitizing, storing, integrating, and analyzing these disparate data sets from both structured and unstructured sources is the major challenge in informatics today.

This is where the burgeoning field of informatics comes into play, and it is being applied in a variety of ways to address the opportunity now present in precision medicine. Though there are many ways in which to talk about

Personalizing Precision Medicine: A Global Voyage from Vision to Reality, First Edition.
Kristin Ciriello Pothier.
© 2017 John Wiley & Sons, Inc. Published 2017 by John Wiley & Sons, Inc.

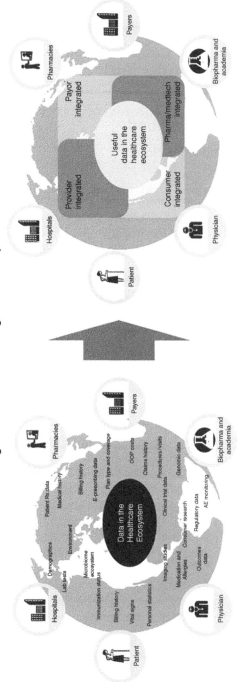

Figure 10.1 Big data stakeholders in the global healthcare ecosystem [1].

informatics, one is to first think of the algorithms powering these diagnostics into one of three very broad categories:

1) Chemoinformatic algorithms, those that analyze the correlations between molecular structure and function at the chemical level
2) Genomic/transcriptomic algorithms, those that specialize in parsing through the genetic code in order to identify abnormalities (e.g., mutations) that may be correlated with heterogeneous diseases like cancer or cystic fibrosis
3) Integrative algorithms, sometimes also rolled into artificial intelligence (AI), which supplement genomic data points with other biological and clinical information for broader predictive power. As Mike Schroepfer, Facebook's chief technology officer, suggested, "The power of AI (artificial intelligence) technology is that it can solve problems that scale to the whole planet" [3], and precision medicine research is certainly no exception. Bioinformatics, chemoinformatics, and integrative informatics all rely heavily on leveraging AI algorithms to decipher reams of raw data and unlock clinically relevant insights. Such models are inherently predictive in nature, relying on statistics, natural language processing (NLP), and other computational techniques to extrapolate outcomes based on the arsenal of data input. Collectively, these aptly named "machine learning" (ML) techniques not only help scientists identify novel biomarkers that may act as druggable targets previously unknown but also help researchers and clinicians segment patients based on *known* targets and therapies. Both are ongoing applications of precision medicine, and this chapter aims to provide a breakdown of how such systems support these efforts across the three segments using genomic sequencing informatics as an example.

Let's Start with What "It" Is

Thanks to technological advances in sequencing technology, it is now possible to sequence a human's entire genome—reading in an individual's entire DNA down to the individual nucleotide base pair level—for less than $1000. Human samples available to researchers and clinicians are heterogeneous, and technological improvements in capture accuracy now allow for relatively small amounts of blood, urine, saliva, and tissue to all be viably sequenced. This has resulted in enormous research opportunity, and hundreds of companies, both large established players like IBM and Google, as well as smaller start-ups, have rushed in to fill the void, each with their own attempts at translating this research opportunity into business potential. This NGS market has ballooned to $2 billion in about 7 years (an active installed base of about 10,000 systems) [4] with some analysts projecting future growth up to $40 billion [5].

In addition to what may be termed the "sequencing revolution," we are now well into the "cloud revolution," as more and more organizations migrate their data (including sequencing data) from their own servers to the cloud (a catch-all term to describe virtual devices that efficiently and securely pool data). This too drives down organizational costs, reducing a barrier to market entry.

Finally, we have been experiencing a proliferation in the number of publically available open-source tools, databases, and ontologies, like reference genomes and the functional annotation languages against which researchers can analyze [6].

The net result of this confluence of factors is that the informatics space has become quite crowded, with market entrants that each offer highly overlapping value propositions, not just within genomics but also the other so-called "-omics" of biological data: proteomics (the study of proteins), metabolomics (the study of metabolites), lipidomics, and also the sea of data having nothing to do at all with science.

Sequencing bioinformatics is currently the primary tool researchers have at their disposal to personalize treatments at more fundamental levels, identifying and targeting both common and rare genetic mutations that are associated with disease. At its most fundamental level, sequencing bioinformatics consists of five key activities, all of which can be time and labor intensive (Figure 10.2):

The basic steps in analyzing big data

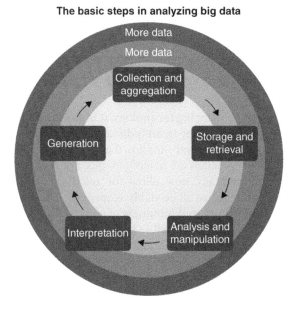

Figure 10.2 Basic steps in analyzing big data. Source: Data from Pothier [1].

1) Properly reading in—or quality checking—the sample
2) Assembling the reads
3) Subsequently identifying variants against the reference data, including insertions/deletions of nucleotides (called "indels") or abnormal number of copies of DNA sections (called "copy number variations")
4) Annotating those variants to predict their functional impact
5) Tying the output together through data visualization and interpretation techniques

The inherent complexity of the process starts with generating the raw sequence, which in turn begins with proper read-in of the sample. In the NGS era, this involves complex preparation measures. As sequencing has evolved, so too have the methods for preparing nucleic acids, but the process is still a long and highly technical one, involving tasks like fragmenting or sizing the target sequences to the desired length, converting target to double-stranded DNA, attaching oligonucleotide adapters to the ends of target fragments, and quantitating the final library product for sequencing [7]. The next step, gene assembly, is also time-intensive, since the sequence is broken down literally into millions of fragments, or short reads. These have to then be reassembled in one of two ways: either by mapping reads to a reference genome *in silico* (e.g., relying on computation as opposed to experimental sample) or putting back together de novo, without the aid of a reference genome. No one method is consistently better than the other (reference assembly is likelier to miss structural variants like chromosomal rearrangements, e.g., whereas de novo assembly fails in highly repetitive regions of the genome), and there are several open-source algorithms and reference libraries available for both.

Once the whole sequence has been generated, researchers may then proceed with discovery-related analyses. After the sequence has been compiled, variants are identified through comparison against reference libraries. Since the human genome is no longer a frontier, variants must be evaluated in the context of preexisting gene annotations [8]. This evaluation requires probabilistic search heuristics that are able to identify both rare and common driver mutations contributing to common diseases with high statistical power and clinical interpretation packages that are able to link those variants to druggable targets.

The last step, visualization and interpretation, is critical in ensuring that output is ultimately actionable, either in a research or clinical sense. There are several visualization and reporting capabilities on the market to display and analyze DNA and RNA sequences, and many of them are free and open source. The main sequencing similarity search algorithm—Basic Local Alignment Search Tool (BLAST)—has been around for more than two decades. In fact, I remember looking up from my bench in my sequencing lab the first time I heard the word from the very first bioinformatics researcher I met, in 1996, simply because I thought it sounded more exciting than it actually was (BLAST!). Yet despite its longevity, even relatively generic output interpreter programs to visualize and select BLAST data are seen as deficient [9]. Notably, bioinformatics solutions that enable efficient sifting through thousands of variants to identify potentially clinically significant subsets are only just now being released [5], and expert manual interpretation of results is often necessary, if not always preferred, a theme not unique to sequencing informatics and one that we will return to later.

Where sequencing bioinformatics addresses drug target discovery at the level of nucleotides, chemoinformatics deals with optimizing antibody–antigen interactions at the structural level. The chemoinformatics market has underpinned fundamental R&D processes for decades now and spans diverse functional areas like medicinal chemistry, pharmacology, and toxicity. And unlike sequencing, chemoinformatics in basic discovery does not require sample: all analysis is performed exclusively *in silico*, or computationally. *In silico* modeling around chemoinformatics involves predicting the chemical and biological properties of compounds given the two- or three-dimensional chemical structure of a molecule [10]. It has proven highly effective in basic discovery well before the advent of personalized medicine, but with the shift in research toward targeted therapies, several types of predictive algorithms have gained traction relating molecular structure to ultimate function within the body, including statistical models such as random forest and artificial neural networks [10].

A currently immature market is also growing for platforms that can not only analyze DNA sequences or molecular interaction effects but also integrate different types of biological and chemical data to construct possible molecular pathways underlying specific diseases in pursuit of new biomarkers or repurposing of existing drugs. Unlike the genome, which does not change over time, an individual's protein and metabolite output captured at any one time is a mere snapshot. Additionally, while the human genome consists of a "mere" 30,000 genes, it has the ability to produce a few million proteins. Analysis of more granular biological signatures is not just a Big Data computing problem, it is a Huge Data problem. Combine this with the fact that proteins, enzymes, and other analytes are more molecularly unstable than nucleotides and thus more difficult to capture with existing lab technology. It may be obvious why this space is so nascent, requiring significant expertise in both computational science and biology. However, cellular processes are inherently dynamic, not static. Hence, genetic sequences may not be enough to identify therapeutic biomarkers, which is why integrative tools may prove to be indispensable in the future for target and drug discovery, particularly in longitudinal studies. In cancer, the capability to analyze the proteomic, metabolomic, and immunological signatures of tumor cells has the potential to transform care from a paradigm that is now very much based on trial-and-error strategies to one that more precisely addresses an individual tumor's molecular profile, profiles that are known to vary greatly between tumors of the same type and even within the same tumor [11]. As an example, one such study found that "although mutations in the *EGFR* gene are uncommon in colon cancer, the EGFR signaling pathway is often elevated" [12].

A common computing approach involves constructing and running large-scale simulations of molecular networks underlying a specific disease in order to predict the effects that certain compounds would have on those pathways. These are concepts rooted in a subdiscipline of science known as systems biology (or network biology), and while this field has been an active area of theoretical research for some time, only recently have scientists been armed with the kind of computational bandwidth needed to conduct validation simulations. Indeed, platforms are taking some truly innovative simulation-based approaches. A team of oncologists and biostatisticians has worked over the past 10 years to complete a digital map of all cancer based on all relevant diagnostic and prognostic variables, including genomic and proteomic data, designed to act as an objective standard to aid in proper diagnosis, treatment, risk assessment, and cost control [13]. In addition to analyzing genetic factors, researchers are also honing in on epigenetic factors of disease (e.g., environmental factors that cause cellular and physiological trait variations that switch genes on and off), evaluating models to predict differential gene expression in cancers like lung and glioblastomas. Such analysis enables researchers to answer questions like "Which is more predictive of lung cancer diagnosis, long-term smoking or the presence of specific genetic mutations?" Several bioinformatics platforms armed with genetic and epigenetic capabilities are now on the market, and they differ more in terms of their relative business utility than by the performance of their underlying algorithms. Today, there is significant variability between systems in terms of total cost of ownership, friendliness of user interface, ability to integrate with other programs, and breadth of additional value-added services (e.g., in-person consultations with a patient's physician), among other nonscientific factors.

Of course, the full power of informatics cannot be harnessed without linking biological output to both scientific literature and outcomes data, allowing researchers to validate and iteratively refine their experimental hypotheses. This clinical data can also generate insights on its own merit. A host of data sources are relevant to this discussion, including literature databases, electronic medical records (EMR/EHR), insurance claims, clinical trials, and real-world evidence like procedural data and doctors' notes. In this world of disparate data, the abilities to efficiently collect from siloed sources, store that data, sanitize and manipulate inputs into a single database, and uncover patterns from this database into a real interpretation are all critical.

In the case of parsing through unstructured data sources like scientific literature and physician notes, an important step is the extraction of concepts and their relationships from a given article, known as information extraction (IE). IE usually begins with named entity recognition (NER), or the correct identification of biomedical terms in free text. Terms might be identified using ontologies (essentially a standardized "vocabulary") or through NLP techniques. An implementation of the latter would involve "tokenizing" text to identify word and sentence boundaries, classifying part of speech (e.g., a noun), and then semantically mapping extracted words to a biomedical category (e.g., a gene or a disease) before a syntactic tree is constructed that represents the structure of the whole sentence or phrase [14]. This is a crucial algorithm when combing through millions of scientific articles to identify and prioritize, say, all relevant experiments around a certain disease associated with the expression of a single gene. Recently, dynamic systems have been developed that can iteratively improve their accuracy over time by comparing discrepancies against manually created "training sets," feeding those discrepancies back into the tool and "learning" from them in subsequent runs. A noteworthy example of a system that specializes in NLP is IBM's Watson supercomputer, originally developed in 2006 to answer Jeopardy! game show questions. As a cognitive system that is able to scale human expertise through the aggregation of millions of data points and proprietary ML algorithms, it is also well equipped to tackle applications in healthcare, and IBM has forged partnerships all over the world for this application. For example, Watson for Oncology ranks treatment options for patients, and Watson Discovery Advisor for Life Sciences identifies correlations among drug candidates. But it isn't all about the science. At the AdvaMed conference in 2016, I was lucky enough to have Roslyn Docktor, director of Watson Health Policy and Middle East & Africa and forger of many of these agreements, to speak on a panel I was running on big data and informatics. She stole the show with her quick wit around my questions and her well-informed views on navigating data worldwide. One of the most interesting observations on the panel, which in addition to IBM included senior executives from Medtronic and ResMed, both device companies pioneering the use of big data and informatics hand in hand with their devices, was that the data being collected is not necessarily all clinical: "The real determinant of our health is data that's created by or about us—our genomes, our behavioral and environment health, but that's only a small percentage. The other 90% is based upon social determinants—where we live, eat, work and play. That data is not found in medical literature." The other panelists followed in agreement, with both discussing the need to parse between nonclinical data and clinical data but the important need for both of them to make a holistic decision about a patient pathway.

Indeed, most senior executives either steeped in bioinformatics or attempting to use it emphasize the partnerships around the medical institutions and other companies who specialize in each component of the data going in are crucial to maintain an output that makes sense to the physician and the patient.

Dr Long Le, the director of Technology Development at MGH Center for Integrated Diagnostics in Boston MA, United States, remarked just this when one of my teams visited him for a tour of his lab and a tour of his brain on the evolution of bioinformatics: "Machine learning will likely empower and enable pathologists to perform more efficiently, particularly for non-challenging, routine cases; this would allow pathology experts to devote more time and focus on the most difficult minority of cases. However, while the machine learning tools are being commoditized, a major challenge is data interoperability: integrating and normalizing different sources (and forms) of data. In addition, there is a need for domain experts to own and drive the data science process during discovery and implementation."

To develop successful informatics tools, Dr Le emphasized that it isn't only about creating the algorithm. The content experts (i.e., the biologists) need to drive hypotheses and insights. These experts need to direct the data scientists developing the tools and infrastructure supporters handling data interoperability along with companies like Google or IBM Watson developing AI.

"In addition to a role like Watson, there're three legs—the biologists, the data scientists, and the infrastructure supporters. You need supervision from the biologists of the data scientists and the infrastructure supporters, and all three legs to support Watson—that's where the success is." At MGH, they are doing just that in Dr Le's lab, ushering a holistic methodology to harness big data and create the informatics tools to really change patient care.

MGH is one of several shining examples of premiere medical institutions harnessing bioinformatics in novel ways. Multiple new companies have also highlighted their specific stake in this space, enhancing IBM Watson's, Google's, and Microsoft's impressive offerings and adding to the bioinformatics arsenal. In 2015 and 2016 alone, hundreds of millions of US dollars have been poured into venture funding of bioinformatics companies, including companies like BlueBee who specialize in cloud-based platforms, Seven Bridges and Syapse who focus on complex scientific and clinical data analysis and interpretation, and Swift and Centrillion Tech who focus on visualization and development support. Outside the United States, there are many others, including companies like Orion Health, a New Zealand-based company who excels at connecting and delivering non-US population sets. These companies, institutions, and others are all partnering with each other to leverage each other's strengths and avoid missteps trying to navigate a space that isn't their own while tackling complex diseases in need of complex analysis. For example, Novartis and Microsoft are collaborating to create AssessMS for multiple sclerosis research, which uses Microsoft's Kinect motion camera and ML software to allow a neurologist to track a patient's movement and assess whether the disease is progressing over time. And Pfizer and IBM Watson are collaborating to build a suite of devices, sensors, and machines that can deliver real-time monitoring of Parkinson's patients to physicians and researchers. These are just two of many collaborations happening in the bioinformatics space, bringing different stake-holders together to provide a holistic solution to researchers and patients. Worldwide, we see this phenomenon continue, as illustrated in the callout box in this chapter.

For those of you who have read this chapter and have a sinking feeling in your stomach on trying to comprehend this space and of all we have left to accomplish, you are not alone. We have not yet fully addressed all of the regulatory challenges associated with using outputs from this mass of data, nor even which regulatory authorities should be involved. We have not solidified AI intellectual property and who really owns bioinformatics. We have not addressed the access and reuse of data that is now abundantly available and not bound by one institutional review board or protected in a locked patient record. We are still early in addressing data security in general. This is of crucial importance, because in the case of biomedical research, the data that can be stolen and misused is not behavioral or economic, but rather the uniquely identifying biological information that fundamentally makes us "us." Already, 29 million individual patient records in the United States alone have been reported compromised in data breaches since 2009, with the number jumping almost 140% from 2012 to 2014 [15]. Data protection needs to catch up to our advancements. According to Marcia Kean, chairman of Strategic Initiatives at Feinstein Kean, chair of the Institute of Medicine's Cancer Informatics Workshop, and an over 30-year veteran in precision medicine, on informatics, "We are driving a Maserati down a winding mountain road in the middle of winter." We have new technology, methodology, and computing power that we never had before, but harnessing it while addressing all of the challenges is what we are still working on and can be a dangerous place to be.

Informatics within the life sciences industry is a market just now starting to come into its own. The market has not been characterized by one or two disruptive technologies, but rather an ever-crowded field of entrants endeavoring to design databases, software, and other technologies to aid clinicians and drug manufacturers in the pursuit of personalized treatment approaches, all jockeying for position. Applications of informatics tools have seemingly as much variation as human biology itself, with some focusing on ligand binding sites, others focusing on the genetic code, and still others focusing on clinical trials literature, to name but a few. What many, if not most, of these next-generation platforms have in common is a reliance on AI and ML concepts, heuristics, and algorithms that can mine vast pools of data to uncover patterns and use those patterns to score variables on their correlation with biological and clinical outcomes. Big tech names have thrown their weight behind these efforts, but even relatively small start-ups have been able to acquire customers around the world, including large hospital systems, diagnostics/testing firms, and even biopharma. With the continued convergence of data science,

Global Spotlight on Informatics

This graphic provides examples of collaboration happening worldwide to bring informatics and big data closer to our physicians and providers (Figure 10.3).

Big data and bioinformatics worldwide

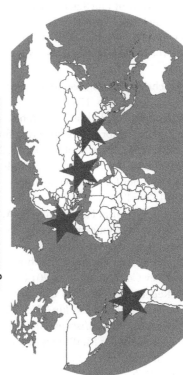

▲ India: Mapmygenome launched Genomepatri Lite to expand its consumer genomics menu to develop an Indian data repository

▲ Europe: $7.6 billion pledged by European genomics researchers to implement the European open science cloud over the next 4 years

▲ Middle East: Tute genomics and Genatak partner to make genomic analysis accessible to clinical laboratories in the Middle East

▲ South America: Fleury partners with IBM Watson Health for Latin America

Figure 10.3 Global collaboration efforts in big data and bioinformatics [1].

computer science, chemistry, molecular biology, and genetics, we may continue to expect innovation to sprout up in unlikely places. This convergence will allow scenarios, like the one Jennifer experienced, to become reality sooner than one might expect.

References

1 Pothier K. The Big Data Debacle in Precision Medicine: Creative Partnering to Usher in a New Age of Bioinformatics Value. Minneapolis, MN: AdvaMed Presentation and Expert Panel; Oct 17, 2016.

2 IDC. The digital universe driving data growth in healthcare. EMC Digital Universe; 2014 [cited Nov 23, 2016]. Available from: http://www.emc.com/analyst-report/digital-universe-healthcare-vertical-report-ar.pdf

3 Clark J. Google chairman thinks AI can help solve world's "hard problems." Bloomberg Technology; Jan 11, 2016 [cited Nov 23, 2016]. Available from: http://www.bloomberg.com/news/articles/2016-01-11/google-chairman-thinks-ai-can-help-solve-world-s-hard-problems-

4 Schenkel D, Rodriguez S, Lin C, Wieschhaus A. Life Science Tool Kit, Seventh Edition. Overview of Life Science Tools Markets and Technologies. Cowen and Company Equity Research; Feb 2015.

5 Groberg J, Iqbal H. Initiation of coverage. Life Sciences Tools, Diagnostics & Genomics: the world is changing, are you ready? UBS Global Research; Mar 9, 2015.

6 Netto GJ, Schrijver I (eds.). Genomic Applications in Pathology. New York: Springer-Verlag; 2014, p. 178.

7 Head SR, Komori HK, LaMere SA, Whisenant T, Van Nieuwerburgh F, Salomon DR, et al. Library construction for next-generation sequencing: overviews and challenges. Biotechniques. 2014;56(2):61.

8 Yandell M, Huff C, Hu H, Singleton M, Moore B, Xing J, et al. A probabilistic disease-gene finder for personal genomes. Genome Res. Sep 2011;21(9):1529–42.

9 Neumann RS, Kumar S, Haverkamp THA, Shalchian-Tabrizi K. BLASTGrabber: a bioinformatic tool for visualization, analysis and sequence selection of massive BLAST data. BMC Bioinf. 2014;15:128.

10 Mitchell JBO. Machine learning methods in chemoinformatics. Wiley Interdiscip Rev Comput Mol Sci. 2014 Sep/Oct;4(5):468–81.

11 Cavallo J. Redefining Cancer, A Conversation With Patrick Soon-Shiong, MD, FRCS(C), FACS. ASCO Post; Jun 25, 2015 [cited Apr 21, 2017]. Available from: http://www.ascopost.com/issues/june-25-2015/redefining-cancer.aspx

12 Li-Pook-Than J, Snyder M. iPOP goes the world: integrated Personalized Omics Profiling and the road towards improved health care. Chem Biol. May 2013;20(5):660–6.

13 COTA Inc. COTA completes digital mapping of all cancers. Nasdaq Global Newswire; Apr 1, 2015 [cited Nov 23, 2016]. Available from: http://globenewswire.com/news-release/2015/04/01/721016/10127192/en/COTA-Completes-Digital-Mapping-of-All-Cancers.html

14 Andronis C, Sharma A, Virvilis V, Deftereos S, Persidis A. Literature mining, ontologies and information visualization for drug repurposing. Brief Bioinform. Jul 2011;12(4):357–68.

15 McCann E. HIPAA data breaches climb 138 percent [Internet]. Healthcare IT News; Feb 6, 2014 [cited Nov 23, 2016]. Available from: http://www.healthcareitnews.com/news/hipaa-data-breaches-climb-138-percent

11

Precision Medicine around the World

India

The newest Wockhardt Hospital emerges out of the crowded and busy streets of Mumbai like a crisp, cool beacon in the surrounding dusty haze and heavy pre-monsoon heat. Wockhardt Hospitals are one of the leading tertiary care/ super specialty healthcare networks in India. Their hospital system overall has a major focus on quality. This newest hospital, modeled after protocols developed by Partners Organization and Massachusetts General Hospital in Boston, MA, US, is only 18 months old and already half-filled with complex surgical patients. The team at Wockhardt Hospitals shared that in their group of hospitals, over 100 quality indicators are measured with a shared quality agenda in which all their clinicians participate. The team also mentioned this hospital is the first in Mumbai to be accredited by the National Accreditation Board for Hospitals and Healthcare Providers (NABH) in India within 18 months of inception, is the first hospital with Leadership in Energy and Environmental Design (LEED) Platinum rating in India, and is certified with Green Business certification (GCBI). Indeed, to maximize their use of space, even the lobby is shared—separated by a wall of glass—with the quality office housing NABH and the US Green Building Council.

But while the hospital system is state of the art by global standards, its access for all Indian citizens, and its growth in offerings and precision medicine for oncology, is not uniform.

The economic success of India over the past decade has resulted in greater purchasing power and increased quality of life for many Indian citizens. As the second largest country in the world, India has a population of 1.25 billion and a total GDP of $1.88 trillion, growing at a rate of 1.3 and 7.3%, respectively [1, 2]. The tremendous growth that India has experienced both in total population and in GDP has driven an increase in demand for all consumer goods, especially healthcare. Healthcare spending in India was $96.3 billion in 2013 and is expected to grow at a rate of 12% through 2018 [3]. India faces many challenges of delivering quality care to rural and urban populations, including accessibility, affordability, and overall disease awareness (Figure 11.1).

Personalizing Precision Medicine: A Global Voyage from Vision to Reality, First Edition.
Kristin Ciriello Pothier.
© 2017 John Wiley & Sons, Inc. Published 2017 by John Wiley & Sons, Inc.

India facts

Figure 11.1 India demographic facts.

Population: 1.25 billion

GDP, 2015: $1.88 trillion

Cancer Prevalence: 3.9 million

Cancer Incidence: 1.1 million

Map:

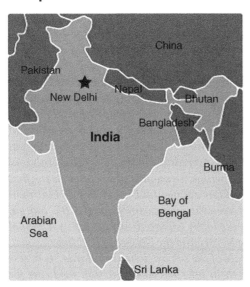

The awareness, according to physician teams I talked with across India, starts with physicians, rather than with the patients as some regions of the world such as the EU and United States are experiencing. In India, there is no empowered patient. Patients still are reliant on all of their medical information to come from the physician rather than from their own research, thus the clinician's own level of awareness is more critical than the patient's awareness in this region at this point. But an important precursor to this is the physician's knowledge of the problem at hand, and even that is challenged in India.

Cancer prevalence (proportion of cases in the population at a given time) in India was reported to be 3.9 million patients in 2015, with a reported incidence (risk of contracting the disease) of 1.1 million [4]. Both prevalence and incidence rate measurements are based on Indian National Cancer Registries, of which there are 27 that are unevenly distributed across India and account for less than 10% of Indian cancer patients [4]. As a result, it is estimated that actual cancer prevalence and incidence could be 1.5–2.0 times larger than the reported [4], leaving a large margin of error in accounting for India's cancer patient population. The variability associated with measurement methodologies makes it difficult to predict and calculate outcomes measurements, but even using the reported numbers, India faces challenges today when it comes to improving outcomes by delivering quality care.

Additionally, for every 1600 diagnosed cancer patients in India, there is only one oncologist. Compared with other more developed countries, it is evident that India is faced with an extreme shortage of qualified physicians to address the cancer patient population. However, a shortage of oncologists is not the only limitation that restricts patient access to quality care. India has only 200–250 comprehensive cancer care centers and institutions that are able to treat a patient throughout the continuum of their treatment. At 0.2 centers per million people, the number of centers pales in comparison with the 4.4 centers per million people available to patients in the United States [4]. India has only 120 PET-CT scanners available for patient screening and assessment, which is only 0.1 per million population compared with 6.2 per million in the United States and 0.9 per million in the United Kingdom [4]. These scanners are critical to diagnosing the type and severity of cancer in order to effectively implement treatment plans. Following diagnosis, there are only about 350 available Linacs, machines that are able to deliver targeted radiation therapy. This limits treatment access to 15–20% of the eligible patient population, compared with international radiation treatment rates of 50–60% [4]. Furthermore, the majority of comprehensive care centers, PET-CT scanners, and Linacs available to screen and treat patients are concentrated in the top eight metropolitan centers, making comprehensive treatment completely inaccessible for patients in rural areas. To accommodate the predicted growth in the number of cancer patients over the next five years, India will have to open 450–500 additional comprehensive care centers that are evenly distributed nationwide. This is an almost impossible task in any country. But by increasing accessibility in this manner, with every center counting, Indian patients will be more likely to be appropriately diagnosed and treated for their conditions.

In 2015, less than 35% of the Indian population had any type of healthcare insurance. Of those covered, the vast majority of plans did not include any type of cancer treatment [4]. The cost of cancer care in India is still markedly lower than the cost of treatment in the United States or the United Kingdom (Figure 11.2) [4], but 75% (200 million) of Indian households have a total annual income that is lower than the average baseline cost of cancer treatment (Figure 11.2) [4]. At this price point, with little or no health insurance coverage,

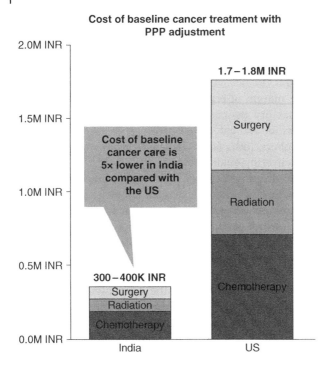

Figure 11.2 Comparison of cost of baseline cancer treatment between India and the United States [5]. Source: Reproduced from EY Report "Call for Action: Expanding Cancer Care in India."

cancer treatment is unavailable to the majority of patients who are diagnosed, forcing many households into debt to pay for the complete treatment. While employees have demanded better health coverage from their employers, and rising income levels will increase disposable income, state and federal governments across India are working to address the lack of healthcare coverage, increasing the percentage of households that will be able to afford the baseline cost of cancer treatment from 31 to 57% by 2020 [4].

A lack of awareness of cancer risk factors and symptoms has driven increased rates of preventable cancers in the developing world. The Indian economy has boomed over the past decade, which has resulted in a shift toward the cancer patient demographics that mirror those in the developed world. Breast, cervical, head and neck, lung, and GI cancers account for more than 60% of solid tumor incidence across female and male cancer patients in India [4]. The high incidence of breast and cervical cancers among women could be prevented by increasing symptom awareness and by advocating for self-examinations and sex education. Head and neck cancers among men are driven by a 25.9% prevalence of smokeless tobacco usage and a 55% increase in alcohol per capita

Figure 11.3 Relative prevalence of smokeless tobacco use and alcohol consumption [5].

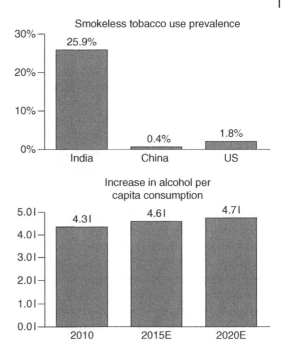

consumption over the past 10 years (Figure 11.3) [4]. These cancers could be prevented by programs that address the dangers of all tobacco use, including smokeless, as well as the risks associated with excessive alcohol consumption. Raising awareness will increase the number of patients that seek preventable care, with the potential impact of decreasing both cancer prevalence and incidence. Admittedly though, awareness campaigns take time, talent, and money, none of which are plentiful in this country.

In addition to increasing awareness of cancer risk factors, India is facing deterioration across other risk factors that further contribute to increased cancer incidence. The problem is getting bigger, not smaller. Not only have tobacco usage and per capita alcohol consumption increased dramatically, but also access to processed foods and poultry has resulted in increases in both obesity and GI cancers. Furthermore, as cities become industrial centers, environmental pollution is contributing to the incidence of lung cancer, with 13 of the top 20 most polluted cities in the world located in India [4, 6]. The Indian demographics are only working to drive increased incidence, as cancer is diagnosed at higher rates in older populations; the aging population will result in an additional 100,000–350,000 newly diagnosed cancer patients annually [4].

In the short term, accessibility, affordability, and awareness of cancer risk factors are key barriers to providing quality cancer care to patients in India. Precision medicine has the capacity to drive the improvement of all three factors if it can be prioritized. By increasing access to molecular and diagnostic testing, preventable cancers that are caused by genetic abnormalities could be addressed at earlier and more curable stages. Currently, only 20–30% of cancers are being diagnosed at stages 1 and 2, which is less than half the diagnosis rate at early stages of China, the United States, and the United Kingdom [4]. Diagnosis at these early stages results in better patient outcomes; the current cancer mortality rate in India is four to six times higher than that of the United States [4]. The use of targeted therapies has been proven to address this problem by improving overall survival, response rates, duration or response, safety, and better quality of life. Precision medicine can also reduce the cost of care by determining the correct treatment modality at the diagnosis phase, reducing the need for changes and adjustments to treatments. Finally, precision medicine used in preventative care as well as cancer-specific care can help to reduce the overall cancer burden in India, helping patients to make more educated decisions about their healthcare consumption. The capacity for precision medicine companies to make a large impact on Indian healthcare has driven a surge in companies that are entering the Indian healthcare market.

Two companies in particular, MedGenome and Strand Life Sciences, are working to integrate precision medicine into cancer care. Both companies provide a comprehensive set of molecular and genomic tests spanning cardiology, endocrinology, ENT, hematology, nephrology, and ophthalmology in addition to oncology testing. The tests are based in NGS sequencing and can be used for familial risk screening as well as solid tumor profiling. MedGenome raised $20 million in series B financing in June 2015, led by VC firm Sequoia Capital, demonstrating that there has been "good progress in the acceptance of genetic testing [in India as] a function of awareness and affordability, both of which have been focus actions for MedGenome as a leader in the market" [7]. Strand Life Sciences has developed partnerships with both large hospital networks and specialist treatment centers in India, establishing a network of key opinion leaders across medical specialties. However, Strand has also focused on "affordability as a key driver of innovation in order to address the Indian market … these tests are not cheap, but they are affordable. The value provided to patients by enabling physicians to make better decisions based on deeper insights justifies the price" [8]. Both companies are working to keep the costs of diagnostic testing low, while also delivering information to physicians that can be used to make better decisions about care for Indian patients. MedGenome and Strand have established themselves in India, and they will have the opportunity to pave the way for further precision medicine companies to enter the space.

In addition, commercial clinical diagnostics reference labs are also enhancing their precision medicine portfolios in India. Metropolis is a commercial reference lab serving India, UAE, Sri Lanka, South Africa, Kenya, Mauritius, and Ghana. It delivers over 30 million tests a year, catering to more than 20,000 laboratories, hospitals, and nursing homes [9]. In a recent visit to their bustling labs in Milind Nagar, Nilesh Shah, Group President, Scientific Services and Operations SBU Head, West India, said, "We have over 4500 tests using 100 technologies and continue to evaluate the most innovative technologies in precision medicine to add to our test menus, including next generation sequencing. A challenge in precision medicine is making sure we are balancing bringing in the newest technology that isn't too expensive for patients to bear while keeping up with the ever changing technology landscape overall."

To that end, India is still working to improve cancer education, access, affordability, and awareness, and there is plenty of opportunity for precision medicine to tackle these issues while simultaneously driving innovation in the field. As additional comprehensive care centers are established and upper-and middle-class out-of-pocket spending drives usage, precision medicine services should continue to enter the market for oncology as well as other indication areas. Increased internet coverage and expanded healthcare registries should only enhance the ability of healthcare providers to utilize new technologies to benefit patients, such as remote monitoring devices or big data analysis to identify and address risk factors within patient populations.

The promise of precision medicine in the Indian market is still in progress, as is many other initiatives to increase the health of the country. While we were being driven through the streets of Delhi, holding on for our dear life as our driver skillfully but breathtakingly navigated around cars, cows, mopeds carrying whole families, and pedestrians all on a 5-lane "highway," we stopped long enough (because of a traffic jam) for me to observe a woman sitting on the side of the highway stringing together a wreath necklace that my driver mentioned she could sell later in the day. Lying next to her, directly in the dirt and surrounded by the heavy dust and gravel kicking up from the highway a foot away from them, was a baby, seemingly taking his morning nap. For once, precision medicine flew out of my head as I wondered what would become of that child, or any of the children that needed to live to adulthood before precision medicine initiatives were even relevant for them. It was an overwhelming thought. But with all that I saw, with the attention now to education and access concentrated in entities like Wockhardt, MedGenome, and Metropolis, among others, the next decade could be a door opener, and precision medicine could be in full force for that generation just getting started now.

References

1 The World Bank. Population growth (annual %). The World Bank; [cited Nov 23, 2016]. Available from: http://data.worldbank.org/indicator/SP.POP.GROW
2 The World Bank. GDP growth (annual %). The World Bank; [cited Nov 23, 2016]. Available from: http://data.worldbank.org/indicator/NY.GDP.MKTP.KD.ZG

3 Dang A, Likhar N, Alok U. Importance of economic evaluation in health care: an Indian perspective. Value Health Reg Issues. 2016;9C:78–83 [cited Apr 11, 2017]. Available from: https://www.ispor.org/policy-perspective_economic-evaluation_India.pdf

4 EY. Call for action: expanding cancer care in India [Internet]. EY; Jul 2015 [cited Nov 23, 2016]. Available from: http://www.ey.com/Publication/vwLUAssets/EY-Call-for-action-expanding-cancer-care-in-india/$FILE/EY-Call-for-action-expanding-cancer-care-in-india.pdf

5 Adapted from: EY. Call for action: expanding cancer care in India [Internet]. EY; Jul 2015 [cited Nov 23, 2016]. Available from: http://www.ey.com/Publication/vwLUAssets/EY-Call-for-action-expanding-cancer-care-in-india/$FILE/EY-Call-for-action-expanding-cancer-care-in-india.pdf

6 World Health Organization. Ambient (outdoor) air pollution in cities database 2014 [Internet]. World Health Organization; 2015 [cited Nov 23, 2016]. Available from: http://www.who.int/phe/health_topics/outdoorair/databases/cities-2014/en/

7 Interview: MedGenome raises series B to advance the practice of precision medicine in India [Internet]. ETHealthWorld; Jul 24, 2015 [cited Nov 23, 2016]. Available from: http://health.economictimes.indiatimes.com/news/diagnostics/interview-medgenome-raises-series-b-to-advance-the-practice-of-precision-medicine-in-india/48193866

8 Nair L. Genetic testing can help in choosing the right drug and the right dose for a patient [Internet]. Express Healthcare; Jun 15, 2015 [cited Nov 23, 2016]. Available from: http://www.expressbpd.com/healthcare/genetic-diagnostics-special/genetic-testing-can-help-in-choosing-the-right-drug-and-the-right-dose-for-a-patient/84958/

9 Metropolis; [cited Nov 23, 2016]. Available from: www.metropolisindia.com

Part 3

The Future

12

A Personalized Stomach

Precision Medicine beyond Cancer

When people talk about precision medicine, they typically talk about the progress being made in cancer, for obvious reasons. But the next frontiers in precision medicine are moving outside of cancer and into other areas of interest to almost everyone: the gut and the head. And when people talk about precision medicine, it is typically in the context of its administration and control in a medical setting. To continue to break the mold, we are going to talk about these new frontiers in and out of the medical setting. The next few chapters delve into our guts, our brains, and our control of precision medicine not only as a patient but also as a consumer.

Let's start with our guts. When I go to different parts of the world, one of my greatest concerns is not the flying or the language or the different cultures or the time differences; it's getting sick. To me, there is nothing worse than having to step onto a podium or into a meeting not knowing whether I can finish the presentation without running to the bathroom to vomit or have diarrhea. It's uncomfortable for me to have to refuse local cuisine because I am just not sure I can handle it or to brush my teeth with bottled water even at the best rated hotels on the planet "just in case." But without fail, the few times I or my colleagues have let down their guards, I see an empty place at the meeting the next morning, or pale, shaky faces as we try to get through the day. And even if it doesn't happen to us, we are joking about it or thinking about it or altering our behavior to protect ourselves. The trip to India I mentioned in the last chapter resulted in not only zero sickness for me but also a loss of 5 pounds and a bigger loss of culinary experiences as I ate crackers and energy bars while everyone else ate real food! Why does this happen to us? Are places in the world so filled with bacteria that we have a public health disaster? How do the people who actually live in those places live there? Are we all that different?

In fact, we are. Or rather, our microbiomes are. A "microbiome" is a community of microorganisms that reside in a specified environmental niche.

Personalizing Precision Medicine: A Global Voyage from Vision to Reality, First Edition.
Kristin Ciriello Pothier.
© 2017 John Wiley & Sons, Inc. Published 2017 by John Wiley & Sons, Inc.

The human microbiome is the collective communities of microbes that call your body home. These microbes are tiny—invisible to the naked eye—and their importance is often overlooked. They are everywhere. Particular regions of the body such as the gut, skin, nose, and mouth provide excellent environments in which diverse subcommunities of microbes can live, thrive, and interact with cell and organ systems (Figure 12.1).

Recent and ongoing medical research is uncovering the role that these microbiomes play in human health. Greater understanding of how communities of microorganisms within different organ systems affect health and even contribute to certain sicknesses and diseases is opening doors to new, perhaps more efficacious therapies than those that are available today. And while the precise nature of how and to what extent the microbiome can positively and negatively affect our health is still being elucidated, the research to date makes one fact abundantly clear: the microbiome and changes in the makeup of microbial communities within the body can directly affect an individual's health.

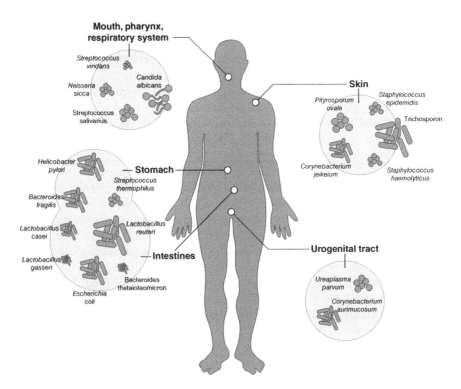

Figure 12.1 Examples of microbe species that live within the human body.

Did You Know?

The inside of your stomach is as unique as your fingerprints. Researchers contributing to the Human Microbiome Project (HMP) are claiming just that. NIH's HMP started in 2008, with the mission of "generating resources that would enable the comprehensive characterization of the human microbiome and analysis of its role in human health and disease." Said another way, researchers are investigating everything and anything living in and on our bodies, whether it be bacteria, viruses, or any other type of microorganism. To give an example of the impact of these microorganisms, there are about 2–6 pounds of bacteria in or around every human body, making up only a portion of your microbiome [1, 2].

But before you get out the soap, know that your microbiome is generally not harmful to you and in fact is essential for maintaining health. Researchers are now correlating what they are seeing in these gut microbiome fingerprints to distinct disease states. And companies like Arivale and AOBiome are using the diagnostics developed from this to create better health options for patients, both sick and well. Before long, caregivers, patients, and healthy people will be able to tailor their diets and their lifestyles not only to how they feel but also right to their fingerprints on the inside.

A greater understanding of which microbes comprise the microbiome and the effect that changes in these microbial communities have to human health has been enabled by the advent of new technologies that have allowed scientists to study the microbiome more efficiently. These include the development of scientific kits to isolate DNA from soil or stool samples with greater ease than previous methods in order to sequence them, like those from Thermo Fisher [3]. Use of these new technologies has sparked a rapid growth in research on the human microbiome. Scientists are striving to better understand the specific mechanisms by which our various microbial communities impact our health, and pharmaceutical companies are racing to use this information to develop novel, more precise therapeutics that can treat diseases more effectively than existing treatments [4].

The list of diseases that may be impacted by changes or imbalances in the human microbiome is a long one. Globally, millions of people are affected by common ailments that are associated with imbalances in microbiome flora. For example, the National Institute of Diabetes and Digestive and Kidney Diseases estimates that 60–70 million Americans—about 1 in 5—are affected by a digestive disease [5]. Within this group, there are over 16 million individuals with bowel diseases, which are a collection of serious, highly morbid conditions that include Crohn's disease, ulcerative colitis, and irritable bowel syndrome (IBS). The microbiome in the gut is believed to play a role in all three of these digestive diseases. The microbiome is also thought to play a role in numerous other highly prevalent conditions, such as asthma, diabetes, autoimmune diseases, and skin diseases such as eczema, to name a few [4].

In order to understand the potential role that the microbiome has in human health and disease, it is necessary to understand what the microbiome is and how it interacts with the organs and systems of the human body. All humans are superorganisms. Our bodies are communities made up of human cells as well as a staggering number of microbial stowaways that call our bodies home. Researchers estimate that a given person's microbiome consists of 100 trillion bacteria, which is 10 times the number of human cells [6]. Bacteria just happen to be—in most cases—considerably smaller than human cells, which is why they make up much less of a percentage by weight. In any case, this community of bacteria within the human body is diverse. There are an estimated 500–2000 different species of bacteria that reside in the gut and digestive tract alone.

Depending on the organ system's area of the body, you will find a unique and diverse community of microbes that differ in composition from the community of microbes in another region of the body. Todd Krueger, president of AOBiome, an innovative biotech company whose mission is to transform human health by altering or rebalancing the bacteria in the human body to treat diseases like hypertension and chronic skin conditions like acne [7], calls it "differential bacteriomics," highlighting the fact the body's microbiome is actually fragmented into communities whose compositions differ by "habitat" [8]. Some of the most diverse communities of microbes can be found in the gut, skin, and nasal passages, and they differ in each person and then community to community worldwide.

We have discovered a great deal of information about the nature and composition of these unique microbial communities through advances in genetic sequencing and informatics software. As discussed in previous chapters, genetic sequencing allows us to investigate the complexities of systems like the microbiome on a genomic level. By looking at the genomic makeup of a given microbial community, researchers can gain insights into which species comprise the microbiome in that region of the body as well as the relative concentrations of each species. Scientists believe that it is the relative presence and concentration of specific bacterial species that can mean the difference between health and disease.

One technique for studying the microbiome is called "metagenomics." Metagenomics refers to the study of genetic material directly from a given environment. By characterizing the diversity of genetic material from bacterial samples taken from a specific region of the body, metagenomic sequencing allows scientists to understand the relative distribution of microbes in that region—that is, which bacterial species are more abundant than others [9].

For example, metagenomic sequencing of bacteria present in a fecal sample can provide information about the types of species residing in the colon and the relative amounts in each species that are present. In a recent study published in

Nature, researchers sequenced the gut microbiomes of 124 individuals. The study found 3.3 million unique microbial genes in the sample, which is 150 times the number of genes in the entire human genome [10]. Using statistical techniques, the researchers estimated that there were 1150 unique bacterial species in the gut microbiome samples—and this doesn't even include the unique bacteria that reside in other regions of the body.

Your diet has a significant influence on the makeup of your gut microbiome. Through metagenomic sequencing, researchers have been able to identify groups or clusters of individuals that can be identified based on a particular distribution of bacteria in their gut. Scientists refer to these groups as "enterotypes." Individuals can be sorted into specific enterotypes based on the distribution of metagenomic sequences in their gut bacteria. For example, the Western diet, high in proteins and fat, is associated with a higher relative amount of bacteria called *Bacteroides* enterotype, while diets high in plant fibers are associated with the bacteria *Prevotella* enterotype [11].

Researchers are currently investigating the clinical relevance of these and similar enterotypes to determine if they could be used as predictors for long-term health risks. One such effort underway is the American Gut Project, which seeks to better understand the extent to which dietary choices (e.g., vegan, paleo, etc.) can impact microbiome composition within a population [12]. For now, however, the research is nascent, and the definitions of the various enterotypes are not yet clear enough to be able to tie them to specific clinical outcomes.

The use of antibiotics and other medical treatments such as chemotherapy can also dramatically alter microbiome composition, both short and long term. Research has shown that antibiotic use can lead to a profound disruption of the microbiome, often with unintended and potentially severe consequences for an individual's lifelong health. Multiple studies have shown that antibiotic use causes an immediate decrease in the diversity of gut microbes [13–15]. And although microbial populations in an adult can bounce back—albeit slowly—after a single course of antibiotics, multiple treatments can cause long-lasting changes in composition of an individual's gut microbiome (Figure 12.2) [15].

A 2016 study published in *Nature* found a correlation between early-life antibiotic use, a long-lasting shift in microbiome composition, and, in turn, an increased risk for metabolic and immunological conditions. The study, which analyzed the antibiotic use and microbiome compositions in a sample of 142 Finnish children, found distinct changes in the microbiome associated with antibiotic use, including the depletion of certain bacterial species such as Actinobacteria and an increase in others like Bacteroidetes and Proteobacteria. These shifts in microbiome composition were also found to be associated with a higher risk for developing asthma and antibiotic-associated weight gain. Given their results, the researchers concluded that "without compromising

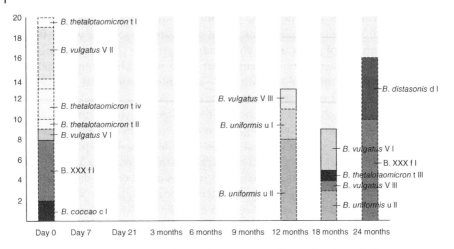

Figure 12.2 Decrease in gut microbiome diversity caused by antibiotic use [13].

clinical practice, the impact of the intestinal microbiota should be considered when prescribing antibiotics" [16].

Finally, and quite relevant to how this chapter began, a gut microbiome can be distinguished on where in the world you live. Research out of Washington University in St. Louis compared the gut microbiomes of 531 people of different ages (from infants to adults, family members, and non-related individuals) living in metropolitan areas of the United States (St. Louis, Philadelphia, and Boulder); Malawi, Africa; and Amazonas, Venezuela. All three areas were of very different socioeconomic, cultural, and geographic makeup. The researchers characterized the bacterial species living in all 531 participants' fecal samples, as well as analyzed 110 gut microbiomes. While some similarities existed in the gut microbiomes of the infants in all three regions, such as genes for basic vitamin biosynthesis and metabolism, the microbiomes in each region had distinct differences, especially occurring between the US microbiomes and the microbiomes of Malawi and Venezuela, although differences also existed between the Malawi and Venezuela microbiomes. Additionally, the bacterial species in the fecal matter of each region were drastically different regardless of age [17]. The differences in the gut microbiomes result in different digestion, absorption, and tolerance of foods in our bodies, and there is early evidence to suggest that travelers with more diversity in their microbiomes are less likely to get sick [18]. So much for all of those well-meaning advice-givers that tell me overcoming sickness while traveling is mind over matter!

While emerging research is increasingly supporting a connection between the microbiome and disease in humans, much of the research to date has shown a correlation rather than establishing a causal link, and a majority of studies have only been conducted on animal models such as mice rather than humans. Still, researchers are unraveling the link between microbiome health and a number of disease states and conditions. For example, long-term shifts in microbiome composition appear to put individuals at a greater risk of autoimmune diseases. Research on mice models has indicated that increased levels of certain types of bacteria can trigger the production of antibodies that attack important cells in the body, causing the body to attack itself. The researchers "demonstrated a link between the microbiome of young mice and the later onset of autoimmune disease" [19].

Other studies also suggest that gastrointestinal microbiome-driven differences in immune system function may also lead to functional abnormalities that produce conditions like asthma and bronchitis. A study comparing children that grew up with household pets with those that grew up without pets indicated that children in pet-free environments were 15% more likely to get bronchitis within the first 2 years of life [20]. Multiple other studies have also begun to establish a link between microbiome composition and an increased risk of inflammatory diseases such as IBS and Crohn's disease, as well as an increased risk of obesity and diabetes due to the microbiome's role in energy homeostasis and body fat storage [21].

Given the growing body of evidence linking microbiome health with disease, it is also important to understand how a healthy microbiome contributes to good health—proper metabolic function, protection from harmful bacteria and other immune functions, and signaling via the so-called "gut–brain" axis [22]. Due to the large genetic diversity of the microbiome, bacteria in the gut can perform a wide range of metabolic functions and synthesize compounds that humans are unable to. For example, bacteria in the gut can produce the complex vitamin cobalamin (e.g., vitamin B12) as well as a variety of different vitamins [23]. By providing biochemical pathways for the metabolism of nondigestible or difficult-to-digest nutrients, the gut microbiome contributes to the recovery of energy and absorbable substrates for the human as well as a supply of energy and nutrients for healthy bacterial growth and proliferation [22]. This role is particularly vital in circumstances where food or nutrition is scarce.

Host protection and development of the immune system is the most studied role of the microbiome today. As discussed previously, microbes exist in habitats in several areas of the body, including the gut, skin, respiratory system, and others. By sheer force in numbers, the microbiome of a healthy person takes up space and uses nutrients that may otherwise be used by pathogenic or disease-causing bacterial species. Researchers call this the "competitive-exclusion" effect [22]. In other words, the balance of bacteria in a healthy microbiome prohibits the growth and proliferation of bacterial species that can hurt the human. Some microbiome bacteria also product antimicrobial substances, called "bacteriocins," that can kill or inhibit other potentially harmful bacterial strains [24].

In addition, a diverse microbiome in the gut can also support the development of a healthy immune system. The intestinal wall is the main interface between the body's immune system and the external environment. Signaling molecules produced by bacteria in the gut bind to receptors in the intestinal wall and cue an immune response. This response includes the release of protective peptides, cytokines, and white blood cells. Proper development of the gut microbiome early in life is believed to play a central role in the immune system later in life [22]. This development is thought to start at birth. Newborns are essentially germ-free for a brief moment at birth but then quickly adopt the microbiome of the mother. A newborn's microbiome composition will even vary depending on the method of delivery. Studies on infants who were delivered via Cesarean section have found that the microbiomes of these infants are more similar to their mother's skin microbiome. Infants born via vaginal delivery have early microbiomes more similar to those of their mother's vaginal and intestinal microbiomes [25].

Throughout the first year of life, an infant's microbiome is characterized by a relatively low level of bacterial species diversity and high rates of flux. Dramatic changes in the microbiome typically occur when there is a shift in feeding mode (e.g., shift from breast milk or formula to solid foods) [10], change in environment (e.g., home to childcare), and genetic and physiological factors (e.g., intestinal pH) that may be unique to that infant [13]. A more stable microbiome composition is reached by around age 3; however, an individual's microbiome composition may still change drastically over the course of that individual's life due to external factors such as diet and use of certain medications such as antibiotics.

Exposure to gut bacteria is also believed to play a role in the prevention of allergies. It is believed that the microbiome stimulates the immune system so that it responds to all potential allergens proportionately. In fact, young children who develop allergies have been found to have a different composition of gut bacteria than those who do not [10].

The last broad function of the microbiome involves a communication system that integrates neural, hormonal, and immunological signals between the gut and the brain. This communication system, called the gut–brain axis, is a two-way system in which the brain commands gastrointestinal functions (e.g., the movement and digestion of food). In the other direction, changes to the composition of the microbiome may contribute to (or compromise) normal healthy brain function, influencing disorders such as anxiety and depression [26]. The clinical consequences of the gut–brain axis, however, are yet to be discovered, and this area continues to be an exciting ongoing research area.

Current Microbiome-Targeted Treatments

Greater understanding of how the microbiome can contribute to human health and disease at a mechanistic level can open the door for more targeted and effective treatments for patients. However, today there are very few treatments on the market that address potential imbalances in the microbiome that are also supported by robust scientific data.

The vast majority of current products aimed at addressing microbiome issues are consumer products, specifically probiotic dietary supplements, for which there is scant clinical evidence of effectiveness. These products claim to supplement "good bacteria" in the gut in order to restore proper balance in the microbiome and, in turn, proper digestive health. The market for these probiotic products is large. It's estimated that the global market for probiotic ingredients, supplements, and food combined was $62.6 billion in 2014 with an annual growth rate of over 7%. The market is continuing to grow rapidly, with some estimates forecasting the total market to reach nearly $100 billion by 2020 [27]. These products come in many forms—pills, drinks, yogurts, and even ice cream. Market growth appears to be driven by rising awareness of probiotics and the role of the microbiome health in general (Figure 12.3) [27].

Between 2011 and 2015, the number of articles on probiotics that were cataloged in the PubMed database tripled, while the number of Google searches related to probiotics increased more than fivefold. We can see evidence of the growing interest in probiotic and microbiome-based products in popular culture as well. Even American actress Jamie Lee Curtis is on television speaking candidly about how a certain probiotic yogurt has made her "regular" again. There is little question that the microbiome and the issue of gut

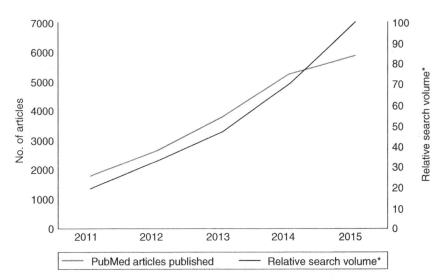

Figure 12.3 Volume of PubMed articles versus Google searches containing term "microbiome" [27]. *Relative Google search volumes with 2015 set at 100%.

health have entered the mainstream. However, it is unclear how effective today's mainstream products are in addressing problems affecting the microbiome.

In short, more research needs to be done on whether most probiotic consumer products are effective. While there is preliminary, largely anecdotal evidence that probiotic products may be helpful in preventing or alleviating certain symptoms (e.g., antibiotic-induced diarrhea), there have been few, if any, clinical studies conducted on probiotic products that have proven their efficacy [28]. Therefore, it is unclear which products may be helpful and which individuals stand to benefit most from their use. A part of this uncertainty is rooted in the fact that, at least in the United States, probiotics are not regulated by agencies like the US Food and Drug Administration (FDA). This is because the FDA currently lacks the authority to establish a formal regulatory category for so-called functional foods, which is the category under which today's probiotic products fall [29]. Consequently, probiotics manufacturers are not required to conduct clinical trials or studies in order to prove the effectiveness of their products, and currently there are no FDA-approved probiotic products on the market.

With that said, simply because a product does not require FDA approval does not mean that its manufacturers can make any health claim that they want. Increasingly, there exists scrutiny around what manufacturers of probiotic products can claim on their product labels. Within the last few years, there have been a number of high-profile class action lawsuits seeking damages for false or unsupported claims by the makers of probiotic supplements. For example, a large company's colon health probiotic supplements were sued for allegations of unsubstantiated clinical claims on its label. In a 2014 class action lawsuit, plaintiffs claimed that scientific evidence was inconclusive or contradicted the company's claims that its products could improve immune system health [30]. As of publication, this case has not been resolved.

As dubious as the clinical claims from probiotics products may be, however, there exists another class of treatments today that has gained much more interest and attention from medical practitioners and the general public alike: fecal transplants. Fecal transplants, which involve the transplantation of one person's fecal matter into another patient's colon, have shown to be effective in treating extreme intestinal infections by restoring a patient's gut with a healthy microbiome. One particular infection that has shown to be both prevented and treated via fecal transplant is *Clostridium difficile*. *C. difficile* is a bacteria that is most commonly acquired by the elderly or patients who have had prolonged treatment with antibiotics, and it is easily spread in the hospital setting. It affects up to half a million Americans each year, with 30,000 deaths as a result [31]. While the infection is generally treated with antibiotics, recurring *C. difficile* infection can be difficult to treat due to the persistence of antibiotic-resistant strains [32]. Some studies have shown cure rates through fecal transplantation of greater than 90% in patients with relapsing *C. difficile* infections [32]. There is also evidence that precision microbial transplantation and reconstitution can lead to increased resistance to *C. difficile* infection [33].

OpenBiome is one example of a company that has placed a bet on the microbial transplantation concept. As one of the first start-ups to successfully bring a medical microbiome-based therapy to market, OpenBiome is a not-for-profit company that collects and cryogenically preserves a bank of human stool for use in patients with recurrent *C. difficile* infections. As of mid-2015, they had shipped over 5000 "treatments" to patients [31].

Overall, however, fecal transplantation and other targeted microbial transplantation therapies are not commonly used today, despite their high apparent efficacy. Given the route of administration and shortage of appropriate donors, widespread use of fecal and microbial transplants is impractical. Therefore, these treatments are most commonly relegated as late or last-line therapies for conditions that cannot be cured or treated through more conventional means. Still, the high clinical efficacy of these transplants has piqued significant interest in the scientific and medical communities. Scientists are most interested in why they are effective—something that is currently unknown at a mechanistic level. It will be important to understand which specific bacteria and which metabolites or proteins, and in what amounts, are effecting the greatest change. This is the new frontier of microbiome-based treatments. Using tools to understand how imbalances in the microbiome can be corrected, how current effective treatments work at a molecular level, creating therapies that precisely address the medical issue, and determining which patients can benefit most from these patients' treatments are all areas that may lead to a more robust market for microbiome-based treatments in the future.

The Future of Precision Medicine and the Microbiome

The future of the microbiome lies in bridging the gap between better understanding the role of the microbiome in human health and the development of medical therapies that perform better than current treatments. From the standpoint of precision medicine, the microbiome represents an exciting, albeit nascent opportunity. The ability to diagnose a particular imbalance in a patient's microbiome and then administer a treatment targeted to correcting or remediating a specific imbalance is the ultimate goal of precision medicine and the microbiome.

Companies that are developing precision medicine treatments around the microbiome must first unravel how either specific bacterial strains or the relative abundance of certain bacteria affects health and disease. From there, they can begin to develop treatments that are targeted at correcting the disease state caused by these imbalances. Recent advances in genetic sequencing and computational techniques have paved the way for researchers to connect these dots in ways that were either not possible or prohibitively expensive in the past [34].

An example of a new investigative technique is the so-called system approaches to learning more about the functions of the microbiome. Rather than focusing on a particular bacterial strain or molecule, "all levels of biological information are investigated (DNA, RNA, proteins, and metabolites) to capture the functional interactions" occurring in the microbiome [35]. Studying these interactions allows researchers to learn things that couldn't otherwise be assessed by studying any one of these processes in isolation [35]. Therefore, systemic approaches offer a more complete and nuanced picture of the microbiome's role in a particular medical condition. For multifaceted and complex diseases such as inflammatory conditions affecting the gut (e.g., inflammatory bowel disease, ulcerative colitis), such a nuanced view is required in order to pair potential therapies with a particular disruption in the microbiome. And companies are already using systemic and other techniques in order to develop potential treatments for patients [36].

Through these new investigative approaches, several companies have begun to develop potential treatments that target imbalances in the microbiome. These emerging treatments can be roughly bucketed into five categories ranging from system-level approaches that pair lifestyle or behavioral changes to a given individual's microbiome to molecular treatments that use a specific drug or compound to correct a disease state caused by an imbalance of bacteria in the gut or other region of the human body.

Transplantation of Defined Bacterial Communities

Transplantation of defined bacterial communities involves transplanting a particular mix of bacterial strains that can restore a patient's microbiome to a healthy state. This approach differs from treatments like fecal transplantation because they do not involve the transfer of human biological material. Instead, defined communities of carefully selected bacterial species can be grown *in vitro* and subsequently transplanted into human recipients. There exist multiple potential advantages to this approach. First, there is no longer a need to screen for ideal donors, which can be scarce. Additionally, scientists have much more control over the relative abundance of specific bacteria contained in the transplant, something that is not possible with fecal samples.

Two companies that have gained attention for their efforts in this space are Vedanta Biosciences and Seres Therapeutics. Both companies are developing mixes of bacterial strains that when administered to donors can potentially treat and cure gastrointestinal diseases. Vedanta Biosciences, for example, is developing a mix of 17 subspecies of Clostridia bacteria that could be a treatment for Crohn's disease and ulcerative colitis [37]. Similarly, Seres Therapeutics is developing what is calling an "Ecobiotic" drug platform that involves transplantation of a bacterial community that can restore function necessary to shift the microbiome to a healthy state [38].

Both Vedanta and Seres have also garnered the attention of large biopharma and health sciences companies. In 2015, Janssen, Johnson & Johnson's biopharmaceutical arm, paid Vedanta an undisclosed initial fee and as much as $241 million in the future for its lead candidate [37], and Nestle Health Science completed an investment of nearly $70 million in Seres Therapeutics in early 2016 [39]. Neither Vedanta nor Seres, however, has yet advanced a drug to human trials, as the lead candidates of each company are still under preclinical investigation.

Next-Generation Probiotics

Next-generation probiotics are single strains of bacteria that can treat imbalances in the microbiome and restore it to a healthy state. As oral therapies, next-generation probiotics would offer considerably more convenience than more invasive treatments such as fecal transplants, which are often administered via the colon. Additionally, bioengineered probiotics may allow treatments to offer benefits previously not possible with natural strains. These synthetic microbes could potentially reside in the gut and turn off and on depending on state of the microbiome, kicking in when necessary.

A company that is currently pioneering in the next-generation probiotics space is Synlogic. Synlogic is currently developing a bioengineered strain of bacteria that has been "programmed" to carry out specific therapeutic functions when there is an imbalance in the microbiome and lay dormant when there is not [40]. One particular therapeutic area where this may be useful is in inflammatory diseases such as inflammatory bowel disease (IBD). In IBD, this could mean that the probiotic can sit inert until a flare-up, at which point the bacteria would release their therapeutic payload. In early 2016, Synlogic partnered with AbbVie in a multiyear collaboration around IBD treatments, with AbbVie agreeing to take on "the burden of regulatory filings, clinical development and future commercialization," as Synlogic's assets are all still preclinical [41].

Secreted Factor/Metabolites and Other Drug Candidates

Perhaps the most targeted application of precision medicine and the microbiome is focusing on specific molecules or metabolites that are associated with a microbiome imbalance. The theory behind this approach is that a given imbalance in the microbiome translates into one or more bacterial strains being over- or underrepresented in the bacterial mix. These strains produce and secrete distinct molecular compounds that, either in overabundance or insufficient amounts, contribute to a given disease by interacting with receptors in the gut or the immune system. By isolating the specific compounds associated with imbalances in the microbiome, researchers hope to be able to treat the symptoms of the diseases directly.

Second Genome is an example of a company that is seeking to develop therapies around secreted factors and metabolites. Its lead candidate, SGM-1019 for IBD and ulcerative colitis, is targeted toward patients who are unresponsive to first-line treatments and may be able to spare the use of very costly and invasive later-line drugs such as antibody-based therapies that must be administered intravenously and may only be modestly effective [42]. Having completed a double-blind phase I FDA trial and with support from organizations such as Janssen, Second Genome's SGM-1019 represents a promising precision medicine therapy for the microbiome and may represent a game-changing approach to rational drug design should it or a similar drug gain approval in the future [43].

The development of drugs that can target and kill specific bacterial strains while preserving others represents yet another molecular-based therapy that shows promise: bacteriocins. Unlike typical antibiotics, which tend to crudely massacre entire communities of bacteria at a time, a given bacteriocin could be used to target a specific strain of disease-causing bacteria while leaving the remaining microbiome intact. Companies like AvidBiotics are developing bacteriocin-like proteins that can target a particular disease-causing bacterium, such as *C. difficile* [44]. Because they are so specific, it is possible that these compounds could even be used as prophylaxis against *C. difficile* infections—something that is currently neither a safe nor a feasible option with antibiotics or alternative treatments such as fecal transplants.

System-Level Approaches

System-level approaches to the microbiome assume interconnectedness between all the processes within the human body, with a functioning microbiome being only one of them. By using health data that encompasses multiple dimensions including an individual's genome, blood, saliva, gut microbiome, and lifestyle, system-level approaches seek to personalize lifestyle and behavioral recommendations to maximize the health of that individual—the difference, for example, between generic advice such as "eat more vegetables" to more precise recommendations such as "based on your DNA and lab work, you are best suited for a Mediterranean diet" [45] (Figure 12.4).

One company that is offering system-level approaches with a gut microbiome underpinning is Arivale. Arivale was founded by Clayton Lewis, a successful entrepreneur and avid triathlete, and Lee Hood, the inventor of the DNA sequencer. Marketed as a "wellness" resource rather than a treatment for a specific condition or disease, Arivale uses numerous health metrics including information about an individual's microbiome in order to recommend tailored, actionable lifestyle changes that they purport can improve the individual's overall health [45].

Arivale's website describes its program with the following preamble: "No one builds a rocket ship without a blueprint or scales Mt. Everest without a guide. So where's the how-to guide to help optimize your wellness? Inside you, of course. Arivale's program takes an intimate and unprecedented look at some of the critical areas of your body and life—your DNA, blood and saliva, and lifestyle—to create a more complete picture of you and your wellness potential" [46]. The company's goal is like the ultimate goal of precision medicine, the right treatment for the right patient at the right time, but it targets all people, not specifically people who are sick or people who are well. Once a user is on the program, Arivale representatives do a full workup of a participant's genome, gut microbiome, basic blood work, and eating, sleeping, exercise, psychological, and basic living habits and pair them with a personalized coach. Then, by looking at the participant's baseline on all dimensions, Arivale is able to recommend changes to optimize their wellness. Example changes that have been made to participants in their program are wide ranging. One participant was found to have elevated GGT (a liver enzyme) and a "G" allele for the APOC3 (rs5128) gene, which increases risk of high triglyceride levels and which led (along with full medical review) to a diagnosis of fatty liver, a serious but treatable condition that she is now working on. Other findings are as simple as a participant who was getting too much iron, or another whose blood lipid levels were too high, or a third whose needed to redistribute his exercise, eating, and sleeping habits to a better self.

System-level platforms using the microbiome as one of many measurements such as Arivale's may be data-driven and holistic, but they are still few offerings from few companies and are reserved for people who can afford them. Arivale's package is $3499, which includes all testing and follow-up, but this pricing shuts out the majority of the general population who could benefit. There is also little peer-reviewed evidence to suggest that companies like Arivale can benefit patients more than simply following the advice of their personal physician, nutritionist, or fitness trainer all at once. However, as we look to broaden the use of precision medicine outside of oncology and into wellness applications (see the Consumer chapter for more information), companies like Arivale are leading the way, one microbiome at a time.

Opportunities and Ongoing Challenges

The state of clinical research around microbiome-based treatments and measurements is still emerging. To date, there are no FDA-approved precision therapies targeting the microbiome and only a handful that have progressed to human trials, although many more may be on the horizon [36]. Demonstrating clinical efficacy will be the next big milestone for microbiome-based treatments. Today's market is inundated with over-the-counter and in-the-grocery-aisle products with dubious claims of efficacy and a lack of proven data. The future of medical treatments targeted toward correcting imbalances in the microbiome will be rationally designed drugs supported by big data and clinical evidence. That day has not come yet, but numerous companies such as Second Genome, AvidBiotics, Synlogic, and others are working vigorously to achieve this goal.

The vast majority of current efforts have focused on the gut microbiome, targeting conditions such as inflammatory bowel disease, ulcerative colitis, Crohn's, and *C. difficile* infections. In some ways, these conditions represent the low-hanging fruit of potential microbiome therapies, as we know the most about the role of gut bacteria in these conditions. However, the microbiome also plays an important role in other systems, such as the skin, mucosa, and even the brain. In 2016, research emerged examining the link between the gut microbiome and Parkinson's disease (PD). Most PD patients complain of constipation for years before PD neurological symptoms appear. "We have discovered for the first time a biological link between the gut microbiome and PD," said lead researcher Sarkis Mazmanian from the California Institute of Technology. The research team used mice that overexpressed a specific protein (α-synuclein) that aggregates in the body to result in the motor dysfunction inherent in PD and showed that when the gut microbiomes of those mice were wiped out through antibiotic use, their PD motor symptoms decreased. They also showed that by transplanting the human gut microbiota of patients with PD, motor dysfunction in the mice increased. Overall, the researchers found that gut bacteria regulate movement disorders in mice, and their research suggests alterations in the human microbiome could be a risk factor for PD [47]. Significant additional research has to be done, but if this discovery holds from mice to humans, it uncovers a significant potential opportunity for the future development of drugs that target neurological disorders.

The road toward FDA-approved precision medicine drugs targeting the microbiome will not be an easy one. In addition to the significant burden of proof needed to pass clinical trials, microbiome treatments may force regulators to create novel strategies for drug approval and protection. For example, requirements for the clinical trials themselves may need to be reassessed. If a microbiome treatment were individualized, studies designed for large cohorts (e.g., the typical phase III clinical trial) may no longer be the best method by which the safety and efficacy of drugs can be determined [48]. For drug development, protecting intellectual property may also be an issue. How does one patent a therapy that is based on naturally occurring bacteria? [36]. What incentive do companies have to develop innovative microbiome therapies if their products cannot be protected?

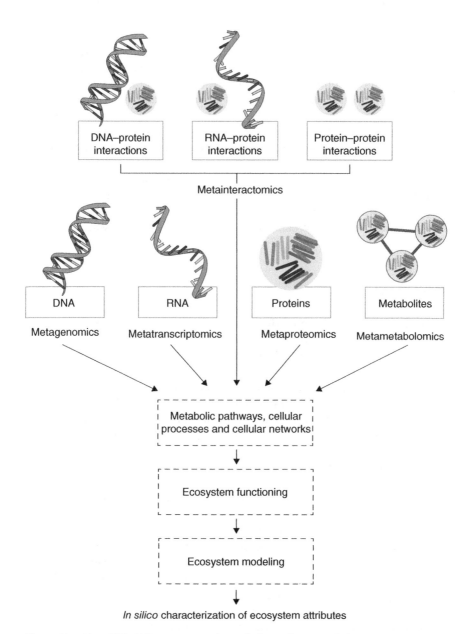

Figure 12.4 How DNA, RNA, proteins, and metabolites influence cellular processes [35].

But potential challenges aside, there is no doubt that researchers have just scratched the surface when it comes to developing potential drugs and medical treatments based around the microbiome. Evidence supporting the connection between a healthy microbiome and the health of numerous systems throughout the body is mounting and becoming increasingly difficult to ignore. The microbiome has also piqued the interests of the general public and is becoming less and less a topic of discussion associated with hippies and health nuts. Today, the microbiome is mainstream. And it's a safe bet that the increasing number of scientific and medical advances around the microbiome will change how we look at health, disease, and precision medicine in the future.

So, as I pack up for my next trip and wish for a personalized vial of Region X microbiome to transform my gut and help me enjoy the region without anxiety, I know I have to wait some time before that wish becomes a reality. In the meantime, my energy bars, my obsession with bottled water, and the almost guarantee of a few pounds lost during the trip will have to do.

References

1 The human microbiome. NIH Human Microbiome Institute; [cited Dec 10, 2016]. Available from: http://hmpdacc.org/overview/about.php
2 Franzosa E. Are Microbiomes the new fingerprints? Harvard Public Health Magazine; 2015 Fall:5.
3 ThermoFisher Scientific. Application note: the human microbiome in 2015 [Internet]. ThermoFisher Scientific; [cited Dec 10, 2016]. Available from: https://www.thermofisher.com/content/dam/LifeTech/global/life-sciences/DNARNAPurification/Files/PG1482-PJ9377-CO210722-App-note-for-Microbiome-launch-Global-FHR.pdf
4 Cho I, Blaser MJ. The human microbiome: at the interface of health and disease. Nat Rev Genet. Mar 13, 2012;13(4):260–70.
5 Digestive diseases statistics for the United States [Internet]. National Institute of Diabetes and Digestive and Kidney Diseases; Nov 2014 [cited Feb 11, 2016]. Available from: https://www.niddk.nih.gov/health-information/health-statistics/Pages/digestive-diseases-statistics-for-the-united-states.aspx
6 Pollan M. Say hello to the 100 trillion bacteria that make up your microbiome. The New York Times Magazine; May 15, 2013 [cited Dec 10, 2016]. Available from: http://www.nytimes.com/2013/05/19/magazine/say-hello-to-the-100-trillion-bacteria-that-make-up-your-microbiome.html
7 AOBiome. [cited Feb 11, 2016]. Available from: https://www.aobiome.com/
8 The Human Microbiome Project Consortium. Structure, function and diversity of the healthy human microbiome. Nature. June 14, 2012;486(7402):207–14.

9 Wang WL, Xu SY, Ren ZG, Tao L, Jiang JW, Zheng SS. Application of metagenomics in the human gut microbiome. World J Gastroenterol. Jan 21, 2015;21(3):803–14.

10 Qin J, Li R, Raes J, Arumugam M, Burgdorf KS, Manichanh C, et al. A human gut microbial gene catalogue established by metagenomic sequencing. Nature. March 4, 2010;464(7285):59–65.

11 Voreades N, Kozil A, Weir TL. Diet and the development of the human intestinal microbiome. Front Microbiol. Sept 22, 2014;5:494.

12 American gut. Human Food Project; [cited Feb 12, 2016]. Available from: http://humanfoodproject.com/americangut/

13 Panda S, El khader I, Casellas F, Vivancos JL, Cors MG, Santiago A, et al. Short-term effect of antibiotics on human gut microbiota. PLoS ONE. Apr 18, 2014;9(4):e95476.

14 Jernberg C, Löfmark S, Edlund C, Jansson JK. Long-term impacts of antibiotic exposure on the human intestinal microbiota. Microbiology. 2010;156(11):3216–23.

15 Goldman B. Repeated antibiotic use alters gut's composition of beneficial microbes, study shows. Stanford Medicine News Center; Sep 13, 2010 [cited Feb 11, 2016]. Available from: https://med.stanford.edu/news/all-news/2010/09/repeated-antibiotic-use-alters-guts-composition-of-beneficial-microbes-study-shows.html

16 Korpela K, Salonen A, Virta LH, Kekkonen RA, Forslund K, Bork P. Intestinal microbiome is related to lifetime antibiotic use in Finnish pre-school children. Nat Commun. Jan 26, 2016;7:10410.

17 Yatsunenko T, Rey FE, Manary MJ, Tehran I, Dominquez-Bello MG, Contreras M, et al. Human gut microbiome viewed across age and geography. Nature. May 9, 2012;486(7402):222–7.

18 Youmans BP, Ajami NJ, Jiang ZD, Campbell F, Wadsworth WD, Petrosino JF, et al. Characterization of the human gut microbiome during travelers' diarrhea. Gut Microbes. 2015;6(2):110–9.

19 Van Praet JT, Donovan E, Vanassche I, Drennan MB, Windels F, Dendooven A, et al. Commensal microbiota influence systemic autoimmune responses. The EMBO Journal. Feb 12, 2015;34(4):466–74.

20 Huang YJ, Boushey HA. The microbiome and asthma. Ann Am Thoracic Soc. Jan 2014;11(Suppl 1):S48–51.

21 Francino MP. Antibiotics and the human gut microbiome: dysbioses and accumulation of resistances. Front Microbiol. Jan 12, 2016;6:1543.

22 Bull MJ, Plummer NT. Part 1: the human gut microbiome in health and disease. Integr Med. Dec 2014;13(6):17–22.

23 LeBlanc JG, Milani C, de Giori GS, Sesma F, van Sinderen D, Ventura M. Bacteria as vitamin suppliers to their host: a gut microbiota perspective. Curr Op Biotechnol. Apr 2013;24(2):160–8.

24 Yang SC, Lin CH, Sung CT, Fang JY. Antibacterial activities of bacteriocins: application in foods and pharmaceuticals. Front Microbiol. May 26, 2014;5:241.

25 Neu J, Rushing J. Cesarean versus vaginal delivery: long term infant outcomes and the hygiene hypothesis. Clin Perinatol. Jun 2011;38(2):321–31.

26 Foster JA, McVey Neufeld KA. Gut-brain axis: how the microbiome influences anxiety and depression. Trends Neurosci. May 2013;36(5):305–12.

27 Driven by rising awareness on gut health, global probiotics market to log 7.40% CAGR from 2014-2020 [Internet]. Transparency Market Research; May 27, 2016 [cited Feb 11, 2016]. Available from: http://www.transparencymarketresearch.com/pressrelease/probiotics-market.htm

28 Probiotics: in depth [Internet]. National Center for Complementary and Integrative Healthcare; 2015 [updated Oct 2016, cited Feb 11, 2016]. Available from: https://nccih.nih.gov/health/probiotics/introduction.htm.

29 Ross S. Functional foods: the food and drug administration perspective. Am J Clin Nutr. Jun 2000;71(6 Suppl):1735s–8s.

30 Top Class Actions. Phillips colon health class action lawsuit survives dismissal [Internet]. Top Class Actions; Nov 10, 2014 [cited Feb 17, 2016]. Available from: http://topclassactions.com/lawsuit-settlements/lawsuit-news/43771-phillips-colon-health-class-action-lawsuit-survives-dismissal/

31 OpenBiome provides its 5000th treatment [Internet]. OpenBiome; Jul 15, 2015 [cited Feb 18, 2016]. Available from: http://www.openbiome.org/press-releases/2015/12/15/openbiome-provides-its-5000th-treatment

32 Rohlke F, Stollman N. Fecal microbiota transplantation in relapsing Clostridium difficile infection. Therap Adv Gastroenterol. Nov 2012;5(6):403–20.

33 Buffie CG, Bucci V, Stein RR, McKenney PT, Ling L, Gobourne A, et al. Precision microbiome reconstitution restores bile acid mediated resistance to Clostridium difficile. Nature Jan 8, 2015;517(7533):205–8.

34 Yong E. Microbiome sequencing offers hope for diagnostics [Internet]. Nature. Mar 23, 2012 [cited Feb 17, 2016]. Available from: http://www.nature.com/news/microbiome-sequencing-offers-hope-for-diagnostics-1.10299

35 Siggins A, Gunnigle E, Abram F. Exploring mixed microbial community functioning: recent advances in metaproteomics. FEMS Microbiol Ecol. May 2012;80(2):265–80.

36 Reardon S. Microbiome therapy gains market traction. Nature. May 13, 2014;509(7500):269–70.

37 Lash A. With Vedanta deal, J&J marks big-pharma milestone in the microbiome [Internet]. Xconomy Exome. Jan 13, 2015 [cited Feb 17, 2016]. Available from: http://www.xconomy.com/boston/2015/01/13/with-vedanta-deal-jj-marks-big-pharma-milestone-in-the-microbiome/

38 Ecobiotic drugs [Internet]. Seres Therapeutics; [cited Mar 3, 2016]. Available from: http://www.serestherapeutics.com/our-science/ecobiotic-drugs

39 Nestlé Health Science completes investment in Seres Health Inc. [Internet]. Nestlé; Jan 6, 2015 [cited Mar 3, 2016]. Available from: http://www.nestle. com/media/news/nestle-health-science-investment-in-seres-health

40 Proprietary platform components [Internet]. Synlogic; [cited Mar 3, 2016]. Available from: http://synlogictx.com/synthetic-biotics/ proprietary-platform-components/

41 Garde D. AbbVie taps Synlogic to take a microbiomic approach to IBD [Internet]. FierceBiotech; Feb 10, 2016 [cited Mar 3, 2016]. Available from: http://www.fiercebiotech.com/story/ abbvie-taps-synlogic-take-microbiomic-approach-ibd/2016-02-10

42 Guidi L, Pugliese D, Armuzzi A. Update on the management of inflammatory bowel disease: specific role of adalimumab. Clin Exp Gastroenterol. 2011;4:163–72.

43 Pipeline [Internet]. Second Genome; [cited Mar 3, 2016]. Available from: http://www.secondgenome.com/development/pipeline/

44 Avidocin^{TM} & Purocin^{TM} Proteins [Internet]. Avid Biotics; [cited Mar 3, 2016]. Available from: http://www.avidbiotics.com/technology/avidocin-proteins/

45 What does being healthy mean to you? [Internet]. Arivale; [cited Mar 3, 2016]. Available from: https://www.arivale.com/how-it-works

46 Home page [Internet]. Arivale; [cited Nov 26, 2016]. Available from: https:// www.arivale.com/

47 Sampson TR, Debelius JW, Thron T, Janssen S, Shastri GG, Ilhan ZE, et al. Gut microbiota regulate motor deficits and neuroinflammation in a model of Parkinson's disease. Cell. Dec 1, 2016;167(6):1469–80.

48 Schork NJ. Personalized medicine: time for one-person trials. Nature. Apr 30, 2015;520(7549):609–11.

13

Consumer Is King

Consumer Applications of Precision Medicine

Jennifer, a healthy nonsmoker in her early 40s who exercises regularly, has noticed that she has been unusually exhausted, queasy, and foggy brained for the past few weeks. Her resting heart rate, blood pressure, and other vitals are continuously being tracked in real time via the latest wearable technology, a wristband that communicates with her primary care physician, immediately flagging and relaying any abnormal readings. She also regularly shares her overall health and well-being with her fitness group online. Although nothing seemed abnormal, Jennifer still worried. Heart disease ran in her family, and she had lost her mother suddenly of a heart attack after her mother had "just felt tired" one day. Her husband had also been feeling queasy for a few weeks now, and she wondered if it perhaps was from the trip they had just taken to India. She also wondered, like some middle-aged women, if she was just losing her mind a little. She texted a message to her doctor, who, upon reading the text, decided to trigger blood testing. Jennifer held her wrist still as the microneedle from the wearable pierced her skin, drew a droplet of blood, retracted back into the wristband, and sent the results back to her doctor. Her doctor, Dr Smith, who happens to be sitting on her outside deck at home, peers at the results from her tablet, raises her eyebrows, and texts Jennifer to come in tomorrow for confirmatory testing.

Upon arrival the next day, another pinprick of blood is drawn and fed into a shoe-box-sized machine right there on the doctor's desk. The device houses a single node in a highly distributed computing grid and is able to leverage supercomputing performance together with a secure cloud-backed architecture to analyze billions of data points from within this single small blood sample in a matter of minutes, including not just genetic information but also Jennifer's entire biological signature, including protein, enzyme, and lipid expression, plus the gathering of her family history and her social media posts on her health and well-being worldwide over the last 3 months.

These data points are systematically compared against "reference models" compiled through population-wide studies to identify relevant biomarkers of interest. These biomarkers are then automatically searched against

Personalizing Precision Medicine: A Global Voyage from Vision to Reality, First Edition.
Kristin Ciriello Pothier.
© 2017 John Wiley & Sons, Inc. Published 2017 by John Wiley & Sons, Inc.

the "literature bank" of the machine, an immense data store of millions of scientific journal articles, electronic health records, and other real-world evidence. Complex association analyses are then performed to generate a ranked list of conditions that Jennifer may have, given her output. When the doctor comes in to talk with Jennifer, she is already armed with the diagnosis, reducing the number of steps that she would have had to do at a first consultation. Her role now is to sit with Jennifer, go through the diagnosis, and counsel her on next steps, which are also given as a list on the report, including treatment recommendations with predictions of which protocols will prove most efficacious for Jennifer, what vitamins or other drugs she should be taking, or whether any of her medications need to be changed. All this information gleaned from the contents of a single blood droplet and billions of pieces of data.

Dr Smith comes in with the report. Other days, she would be diagnosing cancer or heart disease or a neurodegenerative condition. Not today. Jennifer wasn't sick, she was pregnant. Jennifer took a moment to digest the news. She and her husband had hoped for years for a baby, but none had come. Despite the relatively happy news from Jennifer's perspective, at her age, she knew the incidence of chromosomal abnormalities was higher. Also, Jennifer had a sister who had died in childhood from cystic fibrosis, and she didn't know whether she was a carrier. Finally, she wasn't sure whether her multivitamin had enough folate! The doctor was prepared. "All of the tests we could perform on you, and the circulating DNA of the fetus, came back negative. For everything we can test for, including cystic fibrosis and Down Syndrome, the fetus is normal. Based on all of the tests, you are a healthy mother and have a healthy fetus growing inside, and we can now adjust your vitamins, medication, and all else to continue a healthy pregnancy. Would you like to know what you are having?"

"No," says Jennifer, "Let's leave that surprise for another day!" The doctor smiles and folds the page down so that Jennifer cannot see she is having a boy.

Never has healthcare-related information been as accessible to patients as it is today, or it may be in the future. With the advent of the Internet, a plethora of information—from clinical studies, medical journals, diagnostic-focused websites, and even buzzfeed-esque pop websites—is just a short search away. There are clear advantages to this spread of information—for common and easily diagnosable diseases, patients can better understand their symptoms and attempt to treat them before having to secure a visit to a medical professional. At the same time, however, putting this information directly in the hands of patients can easily lead to mistreatment. In fact, sociologists have coined the term "cyberchondria" to describe the phenomena of web-enabled hypochondria [1]—patients' imaginations running wild looking up their symptoms and overdiagnosing their condition based on the information from websites like WebMD, Lifescript, or Wikipedia.

The rise of consumer-based healthcare is shaped by many of the trends in our industry today. As mentioned in previous chapters, we have a worldwide aging population who happens to be aging in an informed, digital world. They are living longer with chronic diseases such as heart disease and diabetes that can be monitored unlike the decades past. The technological advances we are seeing from companies like Apple or Fitbit are expected to impact health outcomes and cost savings in a more meaningful way if used from a healthcare perspective in addition to a consumer perspective (Figure 13.1). Finally, many of our patients today are also Facebook, email, and Internet connected. The rise of the connected consumer, where everyone in a gym spin class hooks up to the teacher's master monitor to have a heart rate contest during class, affects us whether we are sick or well and is shaping innovation from health-based and consumer-based companies worldwide.

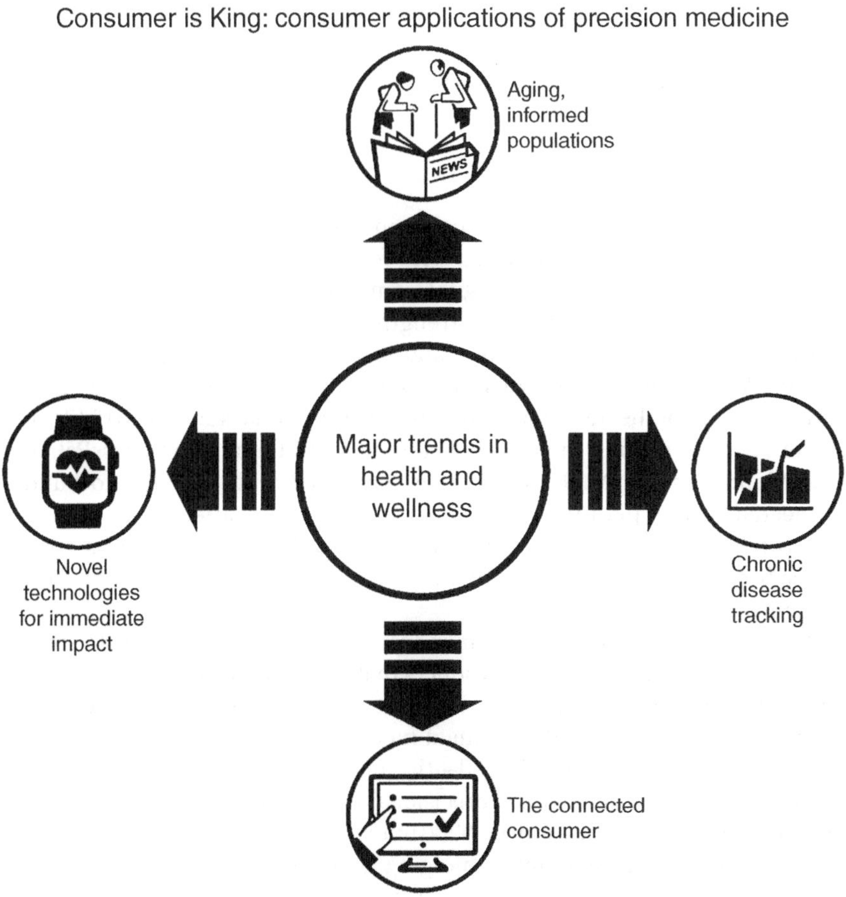

Consumer is King: consumer applications of precision medicine

Figure 13.1 Major trends in health and wellness.

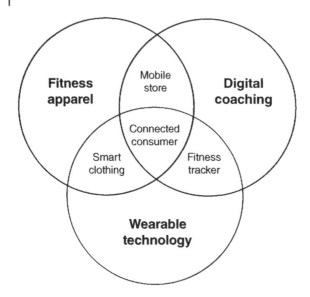

Figure 13.2 Wellness solutions for the connected consumer.

Traditionally consumer-based health and fitness companies are entering into the health and wellness space, with focus on integration of their offerings with digital connectivity. Digital companies are similarly moving into the healthcare space, using their strengths in connectivity to partner with consumer companies. Interesting and sometimes overlapping examples include Apple launching its Healthbook in 2014, Samsung investing in the Series A round of the health scoring-based company dacadoo in 2015 [2], and Under Armour placing mobile applications MapMyRun, MyFitnessPal, and EndoMondo on the Apple watch, plus partnered with IBM [3], to build a connected fitness platform in 2015 and 2016. Fitbit, a leader in wearable fitness tracker technology, acquired Fitstar in 2015, adding a virtual coaching subscription service to provide customized workouts to its loyal consumer base [4].

Over the next decade, the consumer health industry will continue to see the advent of new technologies and products that help patients facilitate meaningful and efficient care at home, self-administered, rather than at a clinic or medical institution administered by trained medical professionals. It's not just the consumers that are being engaged—the investor community has begun to invest heavily in consumer-based precision medicine applications (Figure 13.2). In fact, from 2014 to 2016, over $200 million of venture and private equity investments was invested into emerging companies within the consumer diagnostics and precision medicine industry [5].

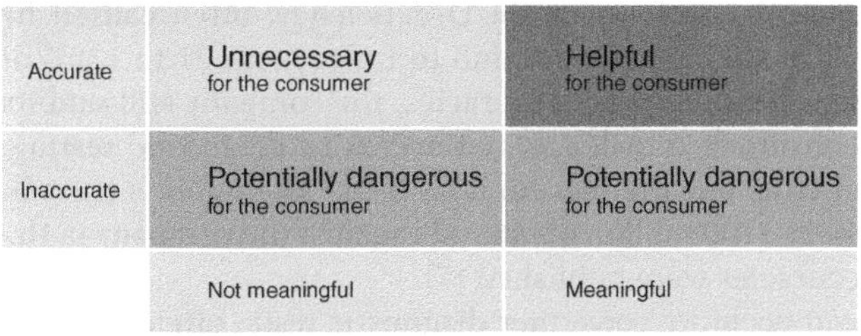

Figure 13.3 Framework for analyzing connected health devices.

For some of us who spend a considerable amount of time in the gym (or in my case thinking about the gym while running through an airport), consumer health applications are a welcome evolution designed to help care for us in our increasingly busy lives. But we can easily be tricked into thinking we are monitoring our health more judiciously than reality, simply by buying into all of the advertising. When we think about these new developments, we really need to do so with two questions in mind:

1) *Is the new development accurate?* Are the results and outputs that are generated a true representation of what is being assessed?
2) *Is the new development clinically meaningful?* Does the outcome of the assessment help the consumer make a decision with regard to their health? (Figure 13.3)

In this chapter, we will explore both of these fundamental elements of consumer precision medicine, starting with accuracy.

Accuracy of Consumer-Based Precision Medicine Products

In order to have a meaningful impact on patients' lives and the lives of those around them, accuracy in consumer health as it relates to precision medicine products and services is essential. Without accuracy, precision medicine has the potential to backfire. In 2016 alone, the medical industry witnessed several consumer-driven diagnostic testing companies that offered promising new technology for patients with questionable accuracy in their products. One diagnostics company promised reduced prices and more convenient testing accessibility for consumers and patients alike, but it turned out that the tests were wildly inaccurate. Upon investigation, it was discovered that nearly 90% of the quality-control checks on their hormone tests were never completed [6]. A different test for vitamin D deficiency, often caused by malnutrition (and lack of sunlight), was found to range from 21 to 130% of the actual levels present. Despite the inaccuracies, the company still said its offerings improved consumer knowledge and access to diagnostic testing, which may have been true, but the inaccuracies broke the public's trust in them and resulted in over $100 million of lawsuits against the company in the months after the inaccuracies were published [7].

When we think about accurate consumer diagnostic tests, often we think about finger-prick blood tests for diabetes or urine tests for pregnancy. While these tests are widely established and have operated as the standard of care for years, there is a whole drive of new precision medicine enabling technologies being brought into or near the home and, more importantly, directly into consumer choice in nonmedical settings. Many of these products are aiming to bridge the gap from "information-centric," nice-to-have technologies to clinically meaningful medical technologies. While we often utilize sensors and scales in our homes to measure weight or track food intake, several additional technologies are being promoted with healthcare aspects today.

One area of particular interest is heart rate monitoring, whose gold standard measurement is the electrocardiogram (ECG). Heart rate monitoring has various applications today—ranging from tracking performance exercise effort to managing physical exertion while pregnant or after a heart attack. For the latter examples, accuracy in heart rate monitoring technology is imperative and always has been. For patients on certain blood pressure medications, doctors will offer guidance to not exceed a certain heart rate during physical activity [8]. There are multiple accurate methods to track heart rates that are less accurate than the ECG but can still be used to physician's satisfaction, ranging from using two fingers next to the thumb and tracking pulses over 10 seconds and multiplying by 6 to more clinical-grade belts that are strapped around a patient's chest [9]. Lately however, we have seen a surge in consumer wearable devices that also claim to monitor heart rate. Large companies transitioning from consumer-friendly technology to consumer healthcare technologies are often challenged in this position to stay within clinical accuracy guidelines, and this came to light in 2016. A large consumer manufacturer, after advertising heart rate monitoring capabilities, was challenged after a study from California State Polytechnic University compared the device's accuracy with the ECG. The study found that the wearable device's readings were inaccurate by up to 20 beats per minute during moderate-to-intense physical activity [10]. For pregnant women or patients on blood pressure medications, these 20 beats per minute could be the difference between comfortable physical exercise and a trip to the emergency room. But for the majority of normal people, this difference is not as severe. Still, the company removed its claim, and its consumer devices remain incredibly popular to this day for more recreational uses.

Clinical Utility of Consumer-Based Precision Medicine Products

The point is clear—accuracy in consumer diagnostics to enable precision medicine is important to many populations. Without accurate devices, using technology to improve consumer awareness of their health and diseases is challenging and sometimes impossible. But after we determine the product is accurate, the next question is, is it meaningful?

Based on the framework earlier, we see that inaccurate products are potentially dangerous and are often not approved for use. And when a test is accurate and meaningful, it can certainly be helpful to the consumer. But the challenge enters when a test or product is accurate, but not meaningful. This circumstance is where unnecessary procedures and costs are created [11].

In the healthcare industry, this meaningfulness is referred to as "clinical utility"—defining whether that information from the product impacts how the consumer or patient could be treated. The concept of clinical utility becomes even more important when considering consumer health and precision medicine, as oftentimes the interpretation and resulting action from the information is even more critical than the accuracy of the tests themselves.

As an example, a well-known debate exists on the clinical utility of a test for prostate-specific antigen (PSA), as it relates to prostate cancer screening [12]. For many years, healthcare organizations have suggested that men begin getting PSA tests as a tool for cancer screening at age 40 or 50. However, recent evidence has suggested that many of these PSA tests resulted in unnecessary long-term effects and harm [13]. The accuracy of the test, as in the number that was being reported by laboratories, was not in question. It was the clinical meaningfulness of the results—upon reading that number, whether the consumer should interpret the results as alarming or as having no impact. To this day, mixed feedback exists on the meaningfulness of the PSA exam. Multiple companies are trying to develop tests that add to or replace PSA with a more clinically meaningful test and are trying to make it even less invasive than a blood sample. For example, Exosome Diagnostics has developed a noninvasive urine test that samples the exosomal RNA in a patient's urine to detect prostate cancer rather than subjecting a man to an invasive biopsy of the prostate. The test ExoDx® Prostate(IntelliScore) enables physicians to predict whether a patient presenting for an initial biopsy does not have high-grade prostate cancer and thus could potentially avoid an initial biopsy and, instead, continue to be monitored [14]. It is one of many noninvasive tests being offered by Exosome. Its turbocharged CEO, my old friend John Boyce, whose track record of nurturing innovation in life sciences goes back over 20 years, pointed to one of their latest press releases for an easy explanation: "Our tests are unique in the liquid biopsy[1] space. By combining the information about disease from exosomes and cell-free DNA captured from any biofluid sample, we are able to achieve the clinical and analytical performance needed for liquid biopsy tests to provide clinicians with real-time, patient specific information that can be used to improve care, select the right therapy, avoid unnecessary procedures and lower overall healthcare costs" [15]. These tests deliver results that can be acted upon from a clinical perspective and also done in a way that is more comfortable for the patient and cost effective for the healthcare system.

Another example of a test with clinical utility received public awareness through actress Angelina Jolie in 2013, mentioned earlier in this book. Jolie is a high-profile US celebrity, known for her leading roles in movies such as Lara Croft: Tomb Raider, Maleficent, and Mr. & Mrs. Smith, in addition to her former Hollywood romance with actor Brad Pitt. Jolie has developed, and continues to have, a wide sphere of influence among the general public, who appreciate her acting ability and many of her charitable and benevolent lifestyle choices [16]. This influence became widely felt among the precision medicine community in 2013 when Jolie announced to the world that she would have a double mastectomy based on the results of her genetic test. To recap, her doctors estimated an 87% risk of developing breast cancer and 50% chance of ovarian cancer over her lifetime based on the results of the test. After weighing the decision with a genetic counselor and considering the fact that her mother had died of breast cancer at age 56, she decided to proactively reduce this risk with the surgical procedure and also made it quite public [17].

The genetic test at the epicenter of this decision was an analysis of the BRCA1 gene, a hereditary gene that is inherited from parents and static throughout one's life [18]. There are several tests available to analyze this gene and determine the results. The risk calculations associated with the test have sufficient clinical utility to enable a person to better understand their health without telling them exactly what to do. The decision and public announcement by Jolie resulted in widespread interest in BRCA1 testing for hereditary risk of both ovarian cancer and breast cancer by women in the United States and abroad. One study in particular indicated that the number of women referred for genetic counseling increased by 90% in 6 months following the announcement [19]. This increase in awareness and motivation to take proactive approaches to cancer management has been positive in the medical community, but only because there is clear clinical utility. Jolie and the medical community

---

1 A "liquid biopsy" is a term used to mean the testing of a body fluid, usually blood, saliva, or urine, which strives to obtain the same clinical diagnostic answer as doing an invasive biopsy of a tissue.

have made it known that this test is not for everyone and that only women with an elevated risk of breast cancer or who have a history of breast cancer in their families should even consider having the BRCA1 test done [20].

This concept of clinical utility has emerged again within the sphere of consumer oncology testing lately, as new offerings have been developed. Recently, new technology has emerged that enables the collection and analysis of circulating tumor DNA (ctDNA) in a patient's blood. While this technology has promise and potential, the current clinical applications are limited, as scientists and companies alike are learning how to best use the technology. In 2016, one particular company tried to claim clinical utility of its ctDNA test, without concrete scientific evidence supporting its use. The test was originally marketed to all patients, where they could submit a tube of blood for analysis for $299–699, and the company would analyze the results and tell the consumer whether tumor ctDNA was present [21]. While the accuracy of the test has not been questioned (the test *does* actually analyze ctDNA), the clinical utility remains questionable. Given the rapid emergence of ctDNA technology, the scientific literature has not found whether certain concentrations or amounts merit additional tests (largely expensive imaging methods and MRIs). When powerful technologies with vast applications are put in the hands of consumers, many of whom are unprepared to interpret the results with the information at hand, the results can be dangerous.

One final area of consumer-based precision medicine has to do with genomics, or the analysis of human genes and DNA as mentioned in previous chapters. While the promise of consumer genomics has been widely touted, the clinical applications of consumer genomics are fairly limited today. Why? Because of a lack of clinical utility. Regardless of how inexpensive the cost of genomic sequencing can become, a large component of genomic testing is still significantly less black and white than some diagnostic tests. Take, for example, consumer genomics companies that are geared toward determining a patient's ancestry and origin. These tests are very clear and have limited clinical or medical actionability—meaning that they're "nice-to-know," interesting conversation starters at cocktail parties, but do not impact disease management. The issue arises when these companies attempt to transition into more clinical applications, such as understanding the risk of developing various diseases, which might include breast cancer and heart disease [22]. Since these assessments require a thorough conversation with a genetic counselor, often unavailable in a DTC setting, companies developing these tests have faced swift regulatory actions from the FDA to better manage the impact that these tests could have on patients. Consumer genomics companies seem to have turned a corner, with several tests now available in a "hybrid lane," focusing on both ancestry-based tests and tests to show carrier status for certain autosomal recessive diseases [23]. These diseases occur when a parent does not have a disease, but there is a chance that the disease could be genetically passed to children. Since the results from this test do not require detailed risk assessments, and have less "gray area," offering these to consumers has been deemed appropriate by many global regulators.

So what happens now? We've identified the importance of having accurate tests offered to consumers. That's taking place. We've identified the need for clinical utility and how it impacts consumer testing. We seem to have turned a corner. Now the final frontier for consumer precision medicine is learning how to manage all of this data. Data management has been touted throughout the precision medicine industry, ranging from the Obama Administration's Precision Medicine Initiative [24] to the National Institutes of Health [25]. While many of these initiatives have focused on the medical aspect of data management, such as clinical genomic information and hospital-based electronic medical records, there are additional data sources that must be incorporated and analyzed to realize the full impact of precision medicine—and that starts and ends with the consumer.

At first glance, it might seem like consumers do not control much data that scientists, pharmaceutical companies, and drug developers would want to understand. However, we need to shift perspective. Advancements in quantum computing and natural language processing have created additional methods for analyzing massive data sets, mentioned in other areas of this book [26]. These data sets have expanded from information manually entered into trackers and journals, which is historically the extent to which consumer data was analyzed. The new data that the healthcare community is interested in lies often at the consumers' fingertips, unlocked through smartphone usage, wearable technologies, and Internet connectivity.

With all of these challenges related to consumer data, it is necessary to understand the industry participants involved in handling and managing all of this data, including upholding consumer privacy. At some point, the consumer bears partial data responsibility, but no more than whether he or she loses their phone in the back of a taxi. The real data management challenges and responsibilities lie with the smartphone manufacturers and software providers. Let's take, for example, the Apple Health application. Built into greater than 15% of all smartphones utilized globally [27], Apple has been a leader in smartphone technology since the late 2000s. From first incorporating cellular service into digital music players, the company has incorporated movement and orientation (sitting and standing) tracking technologies, or accelerometers, into the smartphones to help consumers track their activity and daily steps [28]. While this is the basic offering, the smartphone manufacturer has a more sophisticated platform called ResearchKit and CareKit, which enable its own application developers to create methods to analyze the consumers' data [29]. In fact, within each of these more advanced offerings is a "one-click" option to allow Apple to incorporate and aggregate consumers' data with other smartphone users to glean insights on a population-level basis. These two new platforms have helped to address a challenge in medicine of linking genomic data with patient health and outcomes data. For instance, with "one click," consumers can choose to integrate their movement data and other physical features with their genomic analysis, which was previously completed and integrated with the ResearchKit platform. This creates a very powerful, robust data set that pharmaceutical and medical technology providers are anxious to understand.

While this increase in the amount of data to be analyzed has a massive potential upside for the medical industry, oftentimes consumers question how comfortable they are with this analysis. A 2015 study cited that over 40% of consumers are stressed about smartphone mobile health security [30]. These consumers are concerned that they could be at risk of security data breaches, potentially resulting in disclosures of private health data. Harry Wang, director of Health and Mobile Product at Parks Associates, an IoT market research and consulting organization, cites that "...the connected health industries, device manufacturers, and app developers not only need to ensure they have strong security measures in place but also that consumers are aware of the steps they are taking to protect their data" [31].

Consumer data management is important within healthcare but is equally important in other industries—which will lead to additional incentives to solve problems. With IoT applications ranging from industrial manufacturing to shipping and logistics to drug development, industry experts and regulators are working toward developing the creation of a world with secure data. Now, whether that "smart" data comes from a thermostat, a microwave, a flying drone, or a pacemaker, consumer data, particularly as it relates to healthcare and medicine, is an ever-growing resource that will continue to be mined for value in years to come.

There are many aspects of healthcare and medicine that have been impacted through precision medicine, and perhaps the most important element is consumer engagement and activism. For decades, consumers relied exclusively on the care and recommendations of doctors and healthcare professionals for diagnosis and monitoring of disease. With a push for precision medicine from the government and commercial forces, this enables the consumer to control more elements of their health and well-being than ever before. In order to realize the potential of precision medicine, the consumer needs to have access to accurate tools and understand how to utilize those tools. By addressing accuracy and clinical utility, consumers will be able to more readily approach in-home precision medicine applications. By utilizing in-home precision medicine applications, additional valuable data is generated. By managing this influx of valuable data safely and securely, the healthcare and medical community can learn even more about these patients and help to bring new and innovative products to market. This could be a productive and lucrative cycle, rich with information and insights, but without accuracy, clinical utility, and appropriate data management strategies, precision medicine applications for the consumer may stall while other technology applications outside of healthcare accelerate forward.

Depressing? Somewhat. Realistic? Unfortunately, yes. As much as I enjoy all of the consumer applications coming to market, the lack of regulation and control is a concern if consumers become more dependent on them than their medical professionals, especially if we are really relying on them for credible medical information. According to Jake Orville, president and CEO of the Cleveland HeartLab, one of the most scientifically and clinically grounded cardiovascular diagnostics testing companies in the United States, "Wellness is not a single visit to a physician but rather an on-going and active engagement considering all factors that affect someone's health and well-ness. While the primary point of health engagement should still be a physician, in reality, especially with all of the technology now available to us, many health and wellness decisions will occur outside of a physician's office. Therefore it is up to all of us in the medical diagnostics field to provide proven, credible and meaningful information along the continuum of these interactions." Cleveland HeartLab is doing just that, with a battery of cardiac tests plus a growing wellness menu designed to diagnose and monitor patients/consumers throughout their lives with focus on prevention and maintenance rather than reacting to acute, likely harmful cardiac events. It is a new age of wellness-based diagnostics being monitored with the healthy consumer in mind, as well as the sick patient, and with appropriate precautions, guardrails, and guidance still being put into place, the health-care industry will be better able to leverage and incorporate the consumer for decades of innovation to come, and stories like Jennifer's will become a true and safe reality.

References

1 Loos A. Cyberchondria: too much information for the health anxious patient? J Consumer Health Intern. Oct 2013;17(4):439–45.

2 dacadoo gets funding from Samsung and private investors [Internet]. dacadoo; Jun 3, 2015 [cited Nov 17, 2016]. Available from: https://blog.dacadoo.com/2015/06/03/dacadoo-gets-funding-from-samsung-and-private-investors/

3 Under Armour and IBM to transform personal health and fitness, powered By IBM Watson [Internet]. IBM; Jan 6, 2016 [cited Nov 17, 2016]. Available from: http://www-03.ibm.com/press/us/en/pressrelease/48764.wss

4 Pachal P. Fitbit acquires FitStar for personalized workouts [Internet]. Mashable; Mar 5, 2015 [cited Nov 17, 2016]. Available from: http://mashable.com/2015/03/05/fitbit-acquires-fitstar

5 Analysis based on Pitchbook [Internet]; [cited Nov 17, 2016]. Available from: http://pitchbook.com/

6 Carreyrou J, Weaver C. Theranos devices often failed accuracy requirements [Internet]. Wall Street J; Mar 31, 2016 [cited Nov 17, 2016]. Available from: http://www.wsj.com/articles/theranos-devices-often-failed-accuracy-requirements-1459465578

7 Mukherjee S. Walgreens trashes Theranos in their fiery $140 million lawsuit battle [Internet]. Fortune; Nov 15, 2016 [cited Nov 25, 2016]. Available from: http://fortune.com/2016/11/15/walgreens-theranos-lawsuit-court-documents

8 Your guide to physical activity and your heart [Internet]. National Heart, Lung, and Blood Institute; Jun 2006 [cited Nov 17, 2016]. Available from: http://www.nhlbi.nih.gov/health/resources/heart/obesity-guide-physical-active-html

9 Schönfelder M, Hinterseher G, Peter P, Spitzenpfeil P. Scientific comparison of different online heart rate monitoring systems. Int J Telemed Appl. 2011; 2011:631848.

10 Evenson KR, Goto MM, Furberg RD. Systematic review of the validity and reliability of consumer-wearable activity trackers. Int J Behav Nutr Phys Act. 2015;12:159.

11 Bossuyt PM, Reitsma JB, Linnet K, Moons KG. Beyond diagnostic accuracy: the clinical utility of diagnostic tests. Clin Chem. Dec 2012;58(12):1636–43.

12 PSA—old controversies, new results [Internet]. Harvard Medical School Prostate Knowledge; Jun 2009 [updated May 3, 2011, cited Nov 22, 2016]. Available from: http://www.harvardprostateknowledge.org/psa-old-controversies-new-results

13 Prostate-specific antigen (PSA) test [Internet]. National Cancer Institute; Jul 24, 2012 [cited Nov 17, 2016]. Available from: https://www.cancer.gov/types/prostate/psa-fact-sheet

14 Prostate cancer [Internet]. Exosomedx; [cited Nov 17, 2016]. Available from: http://www.exosomedx.com/prostate-cancer-0

15 Exosome diagnostics announces launch of ExoDx® Prostate(Intelliscore), a completely non-invasive liquid biopsy test to help rule out high-grade prostate cancer [Internet]. Business Wire; Sep 7, 2016 [cited Nov 17, 2016]. Available from:http://www.businesswire.com/news/home/20160907005905/en/Exosome-Diagnostics-Announces-Launch-ExoDx%C2%AE-Prostate-IntelliScore

16 Respect Women. Angelina Jolie: beautifully benevolent [Internet]. Respect Women; Jul 5, 2014 [cited Nov 22, 2016]. Available from: http://respectwomen.co.in/angelina-jolie-beautifully-benevolent/

17 Jolie A. My medical choice [Internet]. The New York Times; May 14, 2013 [cited Nov 18, 2016]. Available from: http://www.nytimes.com/2013/05/14/opinion/my-medical-choice.html

18 BRCA1 & BRCA2 genes: risk for breast & ovarian cancer [Internet]. Memorial Sloan Kettering Cancer Center; [cited Nov 22, 2016]. Available from: https://www.mskcc.org/cancer-care/risk-assessment-screening/hereditary-genetics/genetic-counseling/inherited-risk-breast-ovarian

19 Weintraub A. Angelina Jolie sparks rise in genetic testing for treat breast cancer [Internet]. Oct 19, 2015 [cited Apr 13, 2017]. Available from: http://www.curetoday.com/publications/cure/2015/breast-2015/the-jolie-effect

20 BRCA1 and BRCA2: cancer risk and genetic testing fact sheet [Internet]. National Cancer Institute; Apr 1, 2015 [cited Apr 13, 2017]. Available from: https://www.cancer.gov/about-cancer/causes-prevention/genetics/brca-fact-sheet

21 Pathway Genomics launches first liquid biopsy test to detect cancer-associated mutations in high-risk patients [Internet]. Pathway Genomics; Sep 10, 2015 [cited Nov 17, 2016]. Available from: https://www.pathway.com/pathway-genomics-launches-first-liquid-biopsy-test-to-detect-cancer-associated-mutations-in-high-risk-patients/

22 Lowes R. 23andMe relaunches lower-risk DTC genetic tests [Internet]. Medscape; Oct 30, 2015 [cited Nov 21, 2016]. Available from: http://www.medscape.com/viewarticle/853481

23 FDA permits marketing of first direct-to-consumer genetic carrier test for Bloom syndrome [Internet]. U.S. Food & Drug Administration; Feb 19, 2015 [cited Nov 22, 2016]. Available from: http://www.fda.gov/NewsEvents/Newsroom/PressAnnouncements/ucm435003.htm

24 Bazzoli F. Big IT challenges ahead for precision medicine [Internet]. Health Data Management; Feb 1, 2016 [cited Nov 21, 2016]. Available from: http://www.healthdatamanagement.com/news/big-it-challenges-ahead-for-precision-medicine

25 Landi H. Health leaders talk data analytics, precision medicine and the opportunities, and challenges, for patient care [Internet]. Healthcare Informatics Magazine; Apr 27, 2016 [cited Nov 21, 2016]. Available from: http://www.healthcare-informatics.com/article/health-leaders-talk-data-analytics-precision-medicine-and-opportunities-and-challenges

26 Chandler DL. A new quantum approach to big data [Internet]. MIT News; Jan 25, 2016 [cited Nov 22, 2016]. Available from: http://news.mit.edu/2016/quantum-approach-big-data-0125

27 Smartphone vendor market share, 2016 Q3 [Internet]. IDC; 2016 [cited Nov 21, 2016]. Available from: http://www.idc.com/prodserv/smartphone-market-share.jsp

28 Caddy B. Here's how your phone is tracking you right now [Internet]. TechRadar; Apr 9, 2016 [cited Nov 22, 2016]. Available from: http://www.techradar.com/news/phone-and-communications/mobile-phones/sensory-overload-how-your-smartphone-is-becoming-part-of-you-1210244

29 Maisto M. Apple CareKit, ResearchKit: 6 apps aiming for a healthier world [Internet]. InformationWeek; Mar 7, 2016 [cited Nov 21, 2016]. Available from: http://www.informationweek.com/mobile/apple-carekit-researchkit-6-apps-aiming-for-a-healthier-world/d/d-id/1324852

30 Gruessner V. 41% of consumers stress over smartphone mobile health security. mHealthIntelligence; Sep 2, 2015 [cited Nov 21, 2016]. Available from: http://mhealthintelligence.com/news/41-of-consumers-stress-over-smartphone-mobile-health-security

31 Sprague H, Sternblitz-Rubenstein MS. One-quarter of consumers have privacy concerns about using connected health devices. Parks Associates; Aug 4, 2015 [cited Nov 21, 2016]. Available from: http://www.parksassociates.com/blog/article/pr-aug2015-health-privacy

14

Precision Medicine around the World

The Middle East

The approach from the sky into the Dubai airport is striking. A massive collision of empty sand and expansive sea meets in cool blue and bright beige, and a glittering set of buildings on the shore seems to sit directly on the sand like they were carefully placed there from another world. When walking through the airport and its shopping kiosks and luxe lounges, I was constantly surprised at the newness and lightness of everything around me, considering the long history and perceived challenges of this part of the world. Everything, from the accommodations, to the food and drink, to the souvenirs themselves, seemed rich, copious, and accessible. That changed as soon as I walked out the door and into the 110° heat and reality of both sides of the Middle East.

Recognized for oil products and other exports, the countries of the Middle East vary drastically in reaping benefit from the massive shales that lie underground throughout the region (Figure 14.1). Inequity characterizes the region's resource distribution and, also, its economies. The wealthiest nations—such as Qatar, with its GDP per capita of 132K—rest on vastly different economic and sociopolitical cornerstones than, say, Saudi Arabia, with its GDP per capita of 54K [2]. The former is particularly marked by an almost boundless sense of science's mastery over nature. The United Arab Emirates (UAE), for example, is coping with climate change by building an artificial mountain to maximize rainfall [3]. It is also executing a technique to increase rainfall, called "seeding," where tiny grains of silver, or salt, are shot into existing clouds from planes flying through the updraft of the cloud. These "seeds" then add to the dust that accumulates to produce rain. By early spring of 2013, 47 seeding operations were already underway, firing salt flares into the sky and raining down on what would otherwise be a hot stark desert [4].

The weather isn't the only example of perception differing from reality. Several cancer epidemiologic trends afflict populations in the Middle East where better access to diagnostics and earlier care could improve outcomes. Women in the Middle East who have breast cancer are typically nearly 10 years younger than their counterparts in the United States upon first diagnosis.

Personalizing Precision Medicine: A Global Voyage from Vision to Reality, First Edition.
Kristin Ciriello Pothier.
© 2017 John Wiley & Sons, Inc. Published 2017 by John Wiley & Sons, Inc.

Middle East facts

Definition: 17 countries in Middle East

Population: 420,931,756

Cancer incidence: 538,000
(2012 estimate, IARC Globocan)

Map:

Figure 14.1 Middle East's demographic facts [1].

In addition, colon cancer rates in certain parts of the Middle East are much higher than those in the United States.[1] And oncologists are in short supply. In some countries of the Middle East, 0.4 oncologists exists for every 100,000 people [5, 6]. This number reflects the weighted average number of oncologists

---

1 http://innovatemedtec.com/content/genomics

in Israel, Iraq, Oman, and UAE. At the lowest number per 100,000 highlighted in this book, it speaks to the need for the Middle East to build medical facilities for its population rather than rely on resources out of region, as well as the order of magnitude of work that must be done to provide its population access that rivals regions like the United States and Europe. Access to medical products and care is also complicated by the fact that the population is quite diverse as a total region, and even within a country, there is a different population mix. For example, in the Kingdom of Saudi Arabia (KSA), 67% of the population is "local," or is native to the region, and the remainder is expatriates. But in the UAE, only 11% of the population is local, and the remainder is expatriates [2]. This mix of population by country changes the health insurance makeup and the ability to pay out of pocket for care and influences overall demand for specific precision medicine services.

Until recently, the Middle East also had a substantial reverse medical tourism trend in precision medicine, which is more often talked about than cited. Of the few facts I was able to find publically cited the following:

- In KSA, 55% of patients flown out in 2014 were cancer patients [7].
- In Dubai, oncology was the most sought after medical treatment abroad and was sought by more than one in five of the total overseas patients in Dubai in 2013 [4].
- In Qatar, there was an 11% decrease in patients treated abroad for cancer in 2014 compared with that in 2013 [8].

In talking with physicians in the region, they explained that in all three countries, the patients going abroad were mainly nationals being sponsored by the respective health ministries to seek treatment abroad. However, due to declining oil prices and subsequent budget constraints, including in healthcare, the countries have reduced the number of patients they send abroad for treatment. The practice of outbound medical tourism, hence, is now restricted to a number of affluent nationals who can afford to seek treatments at their own expense, and medical institutions present in the region are developing more healthcare offerings to serve their populations in their home countries.

Perhaps unsurprisingly, then, precision medicine has recently made significant inroads in these same countries as they build up needed healthcare offerings at home. To understand the landscape of precision medicine in the region, three major initiatives are particularly noteworthy: the Catalogue for Transmission Genetics in Arabs, the Saudi Human Genome Program, and the Qatar Genome Project. Dubai's CTGA database is the largest ethnic-based genetic database worldwide, currently hosting a collection of over 1600 records of genetic disorders with their related genes and clinical ramifications. In KSA, the Saudi Human Genome Program, a national research project, studies the genetic basis of all disease both in KSA and throughout the Middle East, with the aim of offering the highest-quality personal care treatments. The Qatar

Genome Project ambitiously aims to provide whole genome sequences for the entire Qatari population as the foundation of personalized healthcare. Its biobank serves as a platform and driver of health research by recruiting large numbers of the Qatari population to contribute biological samples and information about their health and lifestyle [9].

In parallel with these national initiatives are several programs such as Tawam Hospital, which is now offering the first lung cancer screening program in Abu Dhabi and in the UAE through a low-dose computed tomography (LDCT) lung cancer diagnostic screening at its thoracic surgery clinic [10]. In the clinical lab realm, National Reference Laboratory (NRL, a Mubadala company in partnership with LabCorp) operates two state-of-the-art clinical diagnostic laboratories in Abu Dhabi and Dubai and manages multiple others in institutions across the region. Its mission is "to increase the spectrum, coverage and overall efficiency of laboratory testing in the region, while implementing international best practice reference laboratory processes and setting a new benchmark for quality standards" [11]. In addition to continuing to grow their capabilities and offerings in clinical diagnostics, NRL is devoting resources to training and seminars to spread knowledge and quality assurance in order to increase positive experiences with diagnostics to identify patients earlier, with the goal to achieve better clinical outcomes, thus moving forward the potential for precision medicine throughout the region.

Kingdom of Saudi Arabia (KSA)

Highlighting specific regions of the Middle East, let's start with KSA, the largest country in the Middle East at a whopping $2,149,690\,km^2$ but with only 27 million in population. Availability of precision medicine today and access to diagnostic testing in KSA is difficult to measure and obtain public data on. Travels and calls with those in the region, however, allow glimpses into the growing access for the population, especially in providing diagnostic testing. A clinical diagnostics laboratory making significant headway in offering the full gamut of tests for KSA and surrounding regions is Al Borg. Al Borg medical laboratories are the largest chain of private laboratories in the Middle East/North Africa (MENA) with labs in KSA, the UAE, Qatar, Bahrain, Kuwait, Oman, and Ethiopia. When speaking with their management team, their vision exemplified both the challenges and the promise of precision medicine for its people. In addition to its growing menu of tests, Al Borg is also creative in getting patients to and from their laboratories to provide better access to testing. It forged a partnership with Uber for a 2-week period in November of 2015. During that time Al Borg offered all local Uber riders free diabetes screenings. At the same time, Uber was also offering all riders in KSA 20% off for one ride to or from the lab. Dr Sameh El Sheikh, the CEO of

Al Borg says, "At Al Borg, we are committed to our growing menu of diagnostic options for KSA and surrounding regions and have a clear vision for growth that will allow more people access to the diagnostic testing needed for precision medicine." The conversation was honest in identifying the challenges in KSA but also highlighted the energy and the tenacity of a business that continues to serve its region well and wants to make even more rapid progress in the next few years.

While access and offerings are making headway, actual payment for the testing is still a challenge in KSA. Currently, the breakdown of payers for precision medicine testing is varied and dependent on the specific type of test. Patients pay for single-purpose non-genomic diagnostic tests in a number of ways, including out of pocket, private insurance, or institutional reimbursements. Genomic testing is mostly paid for out of pocket but makes up only a small portion of the entire market. Keeping in mind that approximately 14% of the population earns greater than US$100K in household income, out-of-pocket expenses such as genomic tests are certainly accessible but are relatively less affordable for larger swaths of the population compared with those in the UAE, Qatar, and Kuwait. In addition, reimbursement options can further vary depending on citizenship status. While nationals are currently covered for oncology treatment in public hospitals, they are reimbursed for precision medicine tests on a case-by-case basis. On the other hand, non-nationals are covered for oncology treatment based on their insurance class, and they must pay out of pocket for precision medicine tests.

Moving forward, KSA has an opportunity to collaborate with public payers, institutions, and private insurance firms to drive reimbursement of diagnostics and increase access for patients. In the near term, willing oncologists and pathologists can be identified for future collaboration in research and best practice sharing. On the private insurance end, expansion of access will depend on establishing reimbursement precedent and differentiating tests into narrower categories, such as distinguishing hereditary mutations from non-hereditary ones. Collaboration will continue to be key as private insurers recognize the clinical utility of testing, incentivizing reimbursement across a wider spectrum of insurance classes. Near-term wins in the public and institutional payer realm will depend on leveraging international guidelines to promote the usage of tests for precision medicines. Progressive labs like Al Borg can help KSA move swiftly to making precision medicine a reality.

Qatar

The nation of Qatar is a peninsula—100 miles of largely empty Aeolian sand, a tiny thumb attached to KSA—and the few residents are the wealthiest, on average, in the world. So little of the country's land is readily habitable that over

99% of its residents cluster in cities; three-quarters of them live in the coastal capital, Doha, and the surrounding suburbs of Al Rayyan. These residents—just 2.1 million of them—enjoy an average *per capita* GDP of more than $132,000, over 30% higher than that of the world's second richest country, Luxembourg. Unemployment stands at an enviable 0.4% and upward of 14% of households may be millionaires in USD [12]. And the subset of residents who are native Qatari citizens—only 12% of them, since Qatar also experiences the world's highest rate of inbound migration—pay no income tax [1]. Although immigrants often take on lower-status occupations, working as domestic help or to enable the Herculean construction boom in Doha, the country's Gini coefficient (the classic measure of economic inequality) is moderate [13].

Over the past few decades, the proceeds of Qatar's natural resources—oil beneath the blowing sand and natural gas offshore in the Persian Gulf—have underwritten tremendous growth. This income has built Doha into a hyper-modern city, endowed an 11-figure sovereign wealth fund, and won Qatar the 2022 World Cup—with a stadium designed by the late architectural superstar Zaha Hadid. Today, Qatar's resources also have the potential to position the tiny emirate at the forefront of modern medicine. As a compact, wealthy country, Qatar has an opportunity to offer precision medicine to a larger percentage of its citizens than, perhaps, any other nation on Earth. The challenges Qatar still faces, however, are a testament to the deeply human quandaries that are central to the crusade for precision medicine everywhere.

Precision medicine in Qatar began as a vision brought before the country's queen, around 2010. There seem to have been few reasons for concern at the time: oil income was steady, cash was abundant, and the queen proved to be a sympathetic audience. According to Dr Khalid Fakhro, though, the issue was the vision itself. Today, Dr Fakhro is a principal investigator in genetics at the Sidra Medical and Research Center, the institution executing the grand plan defined in 2010. Technical experts like him weren't involved in that plan's gestation, however. "The objectives for Qatar's precision medicine initiatives were conceived of and proposed not by scientists, but by political consultants," Dr Fakhro says. As a result, Sidra has been left to deliver on a loftier goal than most professional geneticists would have advised, or even dreamed of: to sequence the whole genome of every man, woman, and child in Qatar.

The scale, to those in the trade, is breathtaking. Even with today's technology, we live in a world where sequencing 1000 genomes constitutes a major cross-sectional study. The Qatar Genome Programme, as it came to be known, has the highly substantial capacity of 18,000 genomes per year—thanks to multiple X Ten sequencers, industrial-scale behemoths that cost as much as a jet airplane. (Their current sticker price is around $10 million.) Still, at that rate, sequencing the entire population of this rather small country would take more than a century. The Qatar Genome Programme was chartered to undertake a program of unprecedented, almost unimaginable scale. However, as we have seen, it isn't necessary to sequence every possible patient—healthy and otherwise—to have a priceless trove of genomic data. To that end, Dr. Fakhro and his colleagues at the Qatar Genome Programme have carried gamely on. Today, the Programme is linked to the Qatar Biobank for Medical Research—the country's first effort in the world of precision medicine, also founded in 2010. At the Biobank, Qatari residents check in for a 3–4-hour session—free of charge—that includes a comprehensive, leave-no-stone-unturned medical exam with roughly 60 different measurements involved. Patients undergo a full-body MRI; have their body fat, muscle, and bone density measured; and provide samples of blood, urine, and saliva that are analyzed—and then stored. (This is why it is a bio*bank*.) The biobank, in short, provides a continual source of genomic samples, ready for analysis.

The analyses being done today at Sidra are a recent development, though. Here, too, Qatar offers a word of warning for anyone who believes in the promise of precision medicine. For the first 2 years or so of its existence, before Dr Fakhro and his group of colleagues arrived, the Qatar Genome Programme gathered reams of data—and left it on a shelf. There were no analyses at the level of either the population or the individual patient. Qatar had the data—but it didn't have precision medicine.

In this respect, that earlier generation of Sidra was responding to its audience: Qatar's physicians. "For many," according to Dr Fakhro, "…precision medicine is a black box, for which the input is genomic data and the output is therapies targeted and calibrated to an individual patient's condition. Few understand how molecular information can shed light on the pathophysiology of a given cancer." Specialists in rare diseases were an exception. By and large, though, Qatari clinicians didn't know what to ask for and didn't realize what they weren't getting.

Society had, to some extent, created the conditions that allowed this level of incomprehension to survive unchallenged. Qatari physicians are government employees with lifetime job security and, as such, are seen as partisans of the status quo. No aspect of that incentive structure encourages the innovation and continual improvement often seen in the medical systems of other wealthy countries. Notably missing, too, is what Americans would recognize as the loudest voice for improvements in medical care: patients themselves. The Qatari system still does little to encourage patients to be active members of their own care team and to advocate for their own best interests.

In short, precision medicine in Qatar—even today and even in the world's wealthiest per capita country—is still very much in its infancy. Even with a small population and virtually no apparent limit on financial resources—the dramatic opposite, in both respects, of India—there's no simple solution. But if patients today are part of the problem, then tomorrow they may also be part of the solution—the essential part. Individual patients who do their homework, know their disease (or possible disease), demand data, and ask about risks and options are an urgent driver of change in clinical medicine.

United Arab Emirates (UAE)

Considering the UAE's average annual household income relative to other Middle East countries, affordability for precision medicine tests is high. In contrast to KSA, the private insurance firms in the UAE currently reimburse genomic tests irrespective of cancer types based on oncologists' requests. In fact, oncologists have advanced from using only biomarker tests to using more comprehensive genomic tests for treatment selection. Conversely, public institutions remain unwilling to reimburse comprehensive genomic tests and remain out of the picture for other precision medicine tests. The UAE can conclusively be described as the most mature market compared with peer Middle East countries given the high volume of genomics tests prescribed and the deep level of clinical understanding of precision medicine tests by oncologists.

Progression in the UAE's testing market will depend on medical service differentiation and increasing payers' likelihood to adopt through well-documented clinical evidence. Regular scientific sharing between pathologists and oncologists can spur greater understanding of treatment pathways among insurers, building a case for the inclusion of a greater variety of precision medicine tests. According to a supercharged leader at a pharmaceutical company operating in Middle East, "Working with Precision Medicine in Middle East is truly exciting as there is a great interest from physicians and governmental institutions to develop the care for the patients in this direction. That said, there are still hurdles to be overcome in terms of evidence generation, affordability and the shift needed in guidelines and regulations. I, however, believe precision medicine will be a reality in the Middle Eastern key markets in 5-10 years and am proud to be a pioneer and being part of shaping this new market."

Conclusion

The development and access to precision medicine in the Middle East varies just as much as the diversity of its constituent nations. But the growing attention to scientific discovery, the programs across the region, and the recent follow-through development of in-region medical institutions and clinical laboratories to serve the population so they no longer have to leave home are big steps in the right direction. However, for all of the hope and accomplishment displayed in these new medical institutions and laboratories right in front of me, the other side as I mentioned at the beginning of this chapter still remains. There remain challenges with basic living and inequality in the Middle East, especially for women. These observations were, at times, glossed over in conversations with me, an American woman executive coming from a very different upbringing and climate. Indeed, as mere examples discussed with me outside of formal meetings, women in some areas of the Middle East are punished for driving and gender segregation in the workplace is common; in areas where ISIS has control, women and young girls are being sexually abused and sold as slaves, unsanitary and inadequate conditions in refugee camps in areas of the Middle East providing safety to fleeing Syrians are worsening, and there seems a perpetual unrest in some areas of the region so close to those at peace. In the Middle East, where humanitarian and war-related crises are taking such a toll on the region as a whole, complex offerings inherent in precision medicine may take more time to come to fruition compared with other regions of the world.

References

1 Current world population [Internet]. GeoHive; [cited Aug 3, 2016]. Available from: http://www.geohive.com/earth/population_now.aspx

2 The World Factbook [Internet]. Central Intelligence Agency; [cited Aug 3, 2016]. Available from: https://www.cia.gov/library/publications/the-world-factbook/

3 Al Heialy Y. UAE mulls 'man-made mountain' in bid to improve rainfall [Internet]. Arabian Business; May 1, 2016 [cited Aug 3, 2016]. Available from: http://www.arabianbusiness.com/exclusive-uae-mulls-man-made-mountain-in-bid-improve-rainfall-630079.html

4 Butalia N. Cloud seeding: making it rain [Internet]. Khaleej Times; May 8, 2013 [cited Aug 3, 2016]. Available from: http://www.khaleejtimes.com/nation/weather/cloud-seeding-making-it-rain

5 Efrati I. As number of cancer patients in Israel grows, oncology experts on decline. Haaretz [Internet]; Jun 10, 2015 [cited Apr 11, 2017]. Available from: http://www.haaretz.com/israel-news/culture/health/1.660449

6 Parikh P, Mula-Hussain L, Baral R, Ingle P, Narayanan P, Baki M, et al. Afro Middle East Asian Symposium on Cancer Cooperation. South Asian J Cancer [Internet]. Jun 2014;3(2):128–31 [cited Apr 11, 2017]. Available from: https://www.ncbi.nlm.nih.gov/pmc/articles/PMC4014644/table/T3/

7 Ministry of Health [Internet]. Kingdom of Saudi Arabia; [cited Aug 3, 2016]. Available from: http://www.moh.gov.sa/en/pages/default.aspx

8 SCH annual report 2014 [Internet]. Supreme Council of Health, Qatar; Mar 2015 [cited Aug 3, 2016]. Available from: https://www.moph.gov.qa/news/sch-issues-the-annual-report-2014

9 Precision medicine paving the way to better healthcare in Qatar [Internet]; [cited Apr 11, 2017] http://www.internationalinnovation.com/precision-medicine-paving-the-way-to-better-healthcare-in-qatar/

10 Montgomery S. How genomics is making precision medicine possible in the GCC [Internet]. Innovatemedtec; May 3, 2014 [cited Aug 3, 2016]. Available from: http://innovatemedtec.com/content/genomics

11 National Reference Laboratory [Internet]; [cited Aug 3, 2016]. Available from: http://www.nrl.ae/en.html

12 The haves and the have-nots [Internet]. The Economist; Jul 13, 2013 [cited Aug 3, 2016]. Available from: http://www.economist.com/news/special-report/21580630-even-rich-arab-countries-cannot-squander-their-resources-indefinitely-haves-and

13 GINI index (World Bank estimate) [Internet]. The World Bank; [cited Aug 3, 2016]. Available from: http://data.worldbank.org/indicator/SI.POV.GINI

15

Sci-Fi Potential

CRISPR as the Next Novel Frontier in Precision Medicine

The first time I learned about CRISPR was a few years ago, sitting in Stockholm at the Swedish-American Life Science Summit, dying of jet lag and being mildly embarrassed considering I had arrived days before. The Summit is a fantastic collection of about 100 clinical researchers mostly from Europe who come together in a smaller conference almost every year to share their latest insights and, equally important to all participants, socialize with each other outside of the lab. Emmanuelle Charpentier (now the director of the Max Planck Institute for Infection Biology) took the podium and simply wowed all of us with her team's discovery of being able to use a bacterial system called CRISPR/Cas9 to add or delete genes in any type of cell. Minds blown and imaginations running wild, we broke for lunch after her presentation, noting that she immediately got on her cell phone to talk with investors while we inhaled our fish, cucumber, potatoes, and mayonnaise lunch concoction, hoping for some deleting of that as well.

But let's step back a minute. By now the tale of engineering living creatures is steeped into our cultural lexicon. We have Gattaca replaying for decades on our televisions, books like *Brave New World* as requisite school reading, and genetically modified organism (GMO) labels in the grocery stores. But how do we modify people, our animal companions, and our food? Will this help or harm us? What boundaries are being drawn by global governments and technological limitations?

Let's start with square one: what is gene editing? In essence, gene editing is modifying the DNA sequence in a cell—be it plant, animal, or bacteria. As DNA codes for RNA and proteins, modifying the DNA can change the physical makeup of a cell and thus in turn how the cell behaves. Through time, scientists have devised many methods for gene editing. Most of these methods involve "cutting" the DNA and splicing in a desired gene sequence in the cut. Imagine you have the film for a scene in a movie—a man and a woman are walking down the beach, they kiss, she smiles at him, and then we see them walking away from each other. Now, imagine you cut out that smile and replace it with

Personalizing Precision Medicine: A Global Voyage from Vision to Reality, First Edition.
Kristin Ciriello Pothier.
© 2017 John Wiley & Sons, Inc. Published 2017 by John Wiley & Sons, Inc.

her giving him a slap. Most of the scene is the same, but the message is very different, isn't it? Alternatively, imagine you insert a new clip to the scene, where you see that she is wearing a wedding ring but he isn't. Still similar to the original, but it adds more information to the scenario.

This is what gene editing does. It can substitute part of a genetic sequence for another or add an entirely new sequence to a genome. In this chapter we will review different tools for manipulating genes, delve into their applications for human health, and begin to highlight the ethical and political consequences of it all.

For many years, the biggest development in gene editing was the development of zinc finger technology, introduced in the mid-1990s [1]. As with so many biological tools, this technology takes advantage of an existing biological mechanism. All genes in a cell can be turned on and off by a process called transcriptional activation and repression. When a gene is transcriptionally activated, or "turned on," the genetic code (i.e., DNA) is used as the master blueprint to make RNA, and some of these RNAs are used as the blueprint copy to code for proteins. When a gene is transcriptionally repressed, or "turned off," no RNA and therefore no proteins are made from the DNA at that time. This lets a cell respond to the environment dynamically and make what it needs when it needs it.

For this process to work, special proteins need to be able to bind to DNA to turn transcription on and off. A "zinc finger motif" is a particular protein shape that can bind DNA, and scientists have made their own designer versions of zinc fingers that can bind to whatever piece of DNA they need to change (Figure 15.1) [2, 3]. They can then attach this DNA binding tool to a protein piece that cuts DNA, giving them what is called a "zinc finger nuclease" (ZFN), a custom scissor to cut the DNA exactly where they want. In our film analogy, let's think of this as a cut directly after the kiss scene.

This is where another naturally occurring process comes into play. When DNA is damaged, in this case with a cut, the cell has ways of repairing the damage. If scientists put a small piece of DNA, or donor DNA, in the cell at the same time as they add the ZFN, the cell can paste in the donor DNA right where the cut happened. In this case, a new piece of DNA is added, but no DNA is replaced. In our analogy, this is akin to adding the scene of her hand with a ring and his without a ring right after the kiss. Scientists can also put two ZFNs in at once, allowing them to cut out a whole piece of DNA, and provide a donor DNA that replaces the original piece that was cut out, akin to replacing the smile scene with the slap scene. In this case, some DNA is cut out and replaced with donor DNA.

As you might imagine, this was an amazing advance for science, but it isn't without its problems. First, the process is not entirely efficient. When DNA is cut, repairs can happen in various ways, leaving undesired mutations or donor DNA that is improperly pasted in. Another issue is what is called "off-target

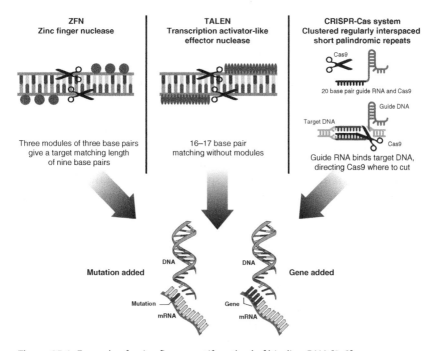

Figure 15.1 Example of a zinc finger motif method of binding DNA [2, 3].

effects" where ZFNs cut the DNA in the wrong place. This is unsurprisingly a common issue in gene editing when the tomato genome, for example, has 900 million base pairs and the human genome has over three billion base pairs. Both of these factors mean that only some cells will be changed in the way scientists want. While this is not ideal for modifying people, it is sufficient in a case in which you are doing research and can test in a petri dish which cells were changed correctly, or, for example, which tomato plant embryos were altered in the right way. Finally, ZFNs can be very expensive, which can also restrict their use.

Another tool scientists devised, a full decade and a half after ZFNs came on the scene, are transcription activator-like (TAL) effector nucleases (TALENs) [4]. These work in a similar way to ZFNs by binding and cutting DNA. Unfortunately, they also are relatively inefficient and face many of the same problems as ZFNs, though they are far easier for scientists to design. Given this, a few companies are currently using this technology to modify genes for potential clinical applications.

The latest advance, coming back to my experience in Sweden, is clustered regularly interspaced short palindromic repeats (CRISPR). CRISPR involves a DNA-cutting protein called Cas9 and a piece of RNA called "guide RNA" that instructs Cas9 where to cut. CRISPR was discovered as part of bacterial defenses against viruses. In the 1980s bacterial genomes were shown to have repeated sequences of unknown significance. Decades later, it was discovered that these repeats matched the DNA code of some viruses and that they were located in the genome very closely to the code for the DNA-cutting Cas protein. By putting these discoveries together, scientists found that bacteria used these repeats that matched viral genes as instructions for where its Cas9 proteins should cut invading viruses and effectively "kill" them. This was a beautiful and simple bacterial immune system defending against viruses. Scientists were able to engineer this natural defense to make a customizable DNA-cutting tool.

The beauty of CRISPR lies in is its facility for customization. Unlike TALENs and ZFNs, which require a scientist to spend weeks or months designing a brand new protein to grab the DNA area of interest, the proteins in CRISPR technology are always the same—a scientist just needs to make a small piece of guide RNA—an easy process that takes just a few days and is very affordable. In essence, ZFNs and TALENs are akin to having to make a brand new scissor and a brand new pattern guide for every piece of fabric you cut, while CRISPR uses the same scissor for each new pattern guide.

Now, how do you get CRISPR/Cas9 into a patient to fix genetic disorders, like cystic fibrosis? To make this work, gene editing technology would need to be delivered to the patient's cells. The most successful way that scientists have been able to insert genes into patient cells is by hijacking a very convenient and ancient tool: viruses. Viruses work by either being engulfed by an unsuspecting cell or attaching to a cell and injecting their DNA or RNA contents into the cell, prompting the cell to create more virus copies. Viruses typically trigger an immune response and can render humans ill. Fortunately, there is a virus called adeno-associated virus (AAV) that scientists have been able to modify to improve its safety and minimize a patient's immune response, yet retain its ability to deliver genetic material into cells. AAV delivery vectors are nonpathogenic, which means they do not cause disease. In addition, they have fantastic staying power and can effectively deliver genes for years.

The problem with AAV has been that viruses are very small and therefore cannot carry much DNA. This historically meant that CRISPR/Cas9 was too big to use this delivery mechanism. Fortunately, scientists have found a smaller version of Cas9 that the virus can deliver [5]. In December 2015, this combination was used to do a proof of concept: scientists used AAV and CRISPR/Cas9 to target the mutated dystrophin gene that causes the disease in Duchenne muscular dystrophy (DMD) mice and through gene editing were able to increase muscle strength throughout the mouse [6]. This was an important victory, as DMD is a serious degenerative muscular disorder condition that affects 1 out of every 5000 newborn males and is caused by a mutation in a single gene [7]. Another method to insert CRISPR/Cas9 involves making the Cas9 DNA-cutting protein and the guide RNA into a complex called a ribonucleoprotein particle (RNP). This premade complex can be mixed with solutions that allow the RNP to enter. This process does not have the ongoing reproductive potential of viruses, but it does allow for easy insertion of CRISPR/Cas9 into cells independent of a complex organism.

Examples of human trials using gene editing include cancer therapies, where immune cells called T cells can be extracted from the blood of cancer patients and then genetically reprogrammed to attack a patient's cancer cells [8, 9]. These cells are called "chimeric antigen receptor" or CAR-T cells (Figure 15.2). Unlike in chemotherapy, which poisons the body's cells—including those that are noncancerous—CAR-T cells are programmed to recognize a specific protein on the surface of cancer cells. In 2016, researchers from West China Hospital used CRISPR technology to fight cancer by removing genes in T cells that impede the cell from attacking the body's own cells. While this might sound counterintuitive, cancerous tumors occur when the body's own cells multiply and build up to unhealthy levels. For a patient who has not responded to chemotherapy and other forms of treatment—both medical and surgical— the removal of the gene that prevents an attack on the body's own cells might be the only solution. However, this gene therapy comes with a serious set of risks. These genetically altered cells could attack healthy cells instead of cancerous cells, trigger a fatal immune response from the body, or produce other deleterious side effects in the patient [11].

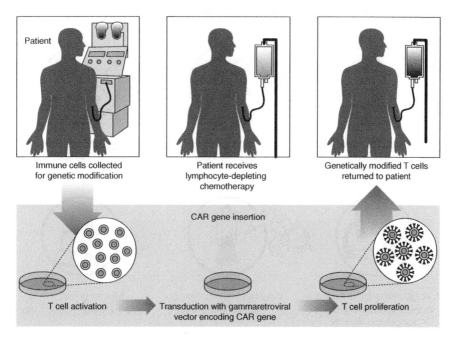

Figure 15.2 CAR gene ion. Source: Adapted from Kochenderfer and Rosenberg [10].

The aforementioned techniques may be useful for people already affected by genetic disease, but the question is, should we prevent the disease from occurring in the first place? Currently, parents who want to have children can get genetic testing to determine if they have a copy of a gene that may make their baby sick. Many of these mutations are recessive, where a parent will have one good copy of the gene and one mutant version and never have any symptoms. This parent is known as a carrier. If this parent has a baby with a non-carrier, their babies will have no risk of developing illness. However, if a carrier has a child with someone who is also a carrier, the couple has a 25% chance that the child will receive both sickness-causing versions of the gene. The American College of Medical Genetics (ACMG) and the American Congress of Obstetricians and Gynecologists (ACOG) periodically review current literature and make recommendations to physicians and genetic counselors on which tests are recommended for which populations; for example, people of Ashkenazi Jewish descent are more likely to have certain inherited genetic variations that may cause illness (e.g., Tay–Sachs disease) [12].

IVF can allow parents to select to implant only fertilized eggs that do not have a particular inherited illness-causing mutation (see Figure 15.3). This is called "preimplantation genetic diagnosis" (PGD). This same process can be used to select embryos with normal chromosomes by "preimplantation genetic screening" (PGS), which, for example, could detect three copies of chromosome 21, which leads to Down syndrome. To carry out PGD/PGS, the

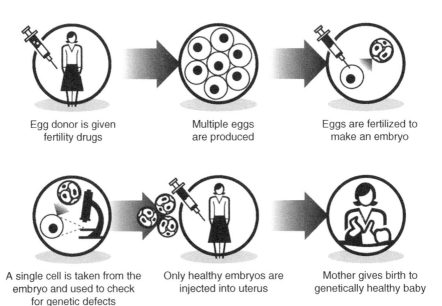

Egg donor is given fertility drugs

Multiple eggs are produced

Eggs are fertilized to make an embryo

A single cell is taken from the embryo and used to check for genetic defects

Only healthy embryos are injected into uterus

Mother gives birth to genetically healthy baby

Figure 15.3 *In vitro* fertilization [24, 25].

physician must grow the fertilized egg until it is a cyst of a few cells and then pull a few cells off to perform genetic testing and determine if it would make a sick child. Only then would a doctor implant the non-carrier eggs. This process is quite expensive, but among those who can afford it, it is common practice to prevent a host of diseases. IVF is legal in the United States and many other countries. Moreover, the International Federation of Fertility Societies noted in 2013 that PGD is allowed in 38 of the 46 countries with specific guidelines and PGS in 26 of the 46 [13]. Some countries such as Germany have strictly limited the application of this technology except in high-risk cases in order to prevent the selection of ideal traits, such as sex [14]. In the United States, it is legal to select the sex of the embryo using these methods.

These methods involve selecting for a favored embryo from those generated by unmodified sperm and eggs. Now, if you were to take this one further, you could get into genetic modification of embryos. Researchers have already begun demonstrating successful proofs of concept for this. For example, in 2015, scientists in China successfully excised the gene myostatin in dogs, which dramatically increased the dogs' strength. This mutation had been discovered in naturally occurring hyper-muscular cows and has been since noted in other species [15]. This success is particularly significant since dogs are often used in biomedical research due to their similarities to humans (in anatomy, physiology, and metabolism). In time, CRISPR may allow scientists to manipulate canine genes to infect dogs with human diseases, which would ultimately enable scientists to conduct potentially lifesaving research that could lead to cures for various fatal diseases [16]. This same technology could also be used to create designer babies by using gene editing to remove genes that cause certain diseases or to add in genes that would produce preferential traits. It is worth noting that we are still far from having gene editing that would work effectively and safely in human embryos.

In 2015, scientists at Sun Yat-sen University in Guangzhou, China, were the first to use CRISPR technology on human embryos. They used CRISPR in nonviable embryos to remove the gene that causes beta-thalassemia, an often fatal blood disorder. Of the 86 embryos, only 28 showed a successful splicing, and only an even smaller fraction showed that CRISPR successfully replaced the altered gene. While this effort was not an enormous success in itself, it is a major advancement in the human application of CRISPR, and it has spurred vigorous ethical debate over the moral implications of altering human embryonic cells. Currently, scientists are working on how to increase the efficacy of the CRISPR technology by improving the targeting mechanism that helps the enzymes determine where the splicing should take place and by altering the way enzymes enter the cells, which impacts the length of the enzyme's lifespan and the buildup of mutations [17].

CRISPR can have great impact on human health even when it is employed on nonhuman subjects. Due to a severe dearth in donated organs both from live donors and from cadavers, people all around the world are dying as they wait on organ transplantation lists. However, using CRISPR technology, researchers at the University of California, Davis, have been working to grow human organs in pigs. So far, the Davis team has used CRISPR to remove the part of the DNA of an embryonic pig that would allow it to develop a pancreas. In its stead, they have placed human induced pluripotent stem cells known as iPS, which can develop into any type of organ cell. By implanting iPS cells, scientists intend for the fetal pig to develop a human pancreas, which could one day be used in a human organ transplant [18]. This incredible advancement in medical technology brings with it various ethical concerns. The growth of human organs in pigs has prompted some fear that the developing pigs might be more humanlike than piglike—an exaggerated and overstated fear [19].

However, once patients get that lifesaving organ transplant, their host immune response can identify the new organ as foreign and potentially dangerous and attack it, prompting the body to reject the organ. However, researchers are using CRISPR to try to develop techniques to prevent organ rejection. Scientists have found that genes that prompt animals to reject foreign organs can be deleted with CRISPR technology—allowing dramatic advancements in xenotransplantation, the transplantation of organs of one animal into another of a different species [19]. Scientists have seen more improvements in the transplantation of individual cells such as islet cells than in entire organs, and they have successfully used gene editing technology to shield pig islet cells with a substance that prevents the pig cells from interfering with the human immune system [19]. This technology is complicated, though, and one concern is that the transplantation of pig and animal organs into humans will bring with it animal diseases as well. In addition, since animals live far shorter lives than humans, another concern is that the transplantation of an animal organ might not last as long as the human lifespan. Even more unfortunate, the use of an animal organ as a holdover until a human organ is available would also prove challenging, as humans grow dense tissue around their new organs—impeding a second transplant surgery [20].

CRISPR technology is improving and scientists are making leaps, especially in nonhuman subjects. Scientists have successfully employed CRISPR to protect pigs from porcine reproductive and respiratory syndrome virus known as PRRSV. By using this gene editing tool, scientists removed a gene in the porcine DNA that allows the virus to enter and infect its subject. Scientists have also used CRISPR to breed cows without horns. This is particularly significant as farmers are often forced to remove their cows' horns, which represent a danger to the farmer. CRISPR protects cows from that painful process by allowing them to be born without horns at all and saves the farmers money [20].

Gene editing can also be employed to create foods that are more nutritious or easier to cultivate in order to prevent famine and diseases of malnourishment. The overwhelming scientific consensus is that genetically modified foods are safe [21]. One classic example is golden rice, a genetically modified food product that contains beta-carotene, a major source of vitamin A. For millions of people around the world, rice is the main component of their diet, and rice gruel is often used for young children as their first solid food. As a result, these young children can experience vitamin A deficiency, which is highly detrimental to their development and overall health. The introduction of golden rice, though at times controversial due to the polemic around genetically modified foods, has provided a valuable source of vitamin A for these developing minds [22]. Historically, genetically modified crops were made by inserting small units of foreign DNA into plants, often by shooting the DNA into plant embryos with what are called "gene guns," using a virus, or using electrical pulses. However, this process is highly inefficient, and gene editing is a much faster method to get desired results.

The use of CRISPR is the next step in plant modification. In 2015, Yinong Yang, a scientist from Pennsylvania State University, successfully employed CRISPR technology to produce a white button mushroom with reduced browning. Yang altered the mushroom's genetic code by removing the six genes that produce the enzyme polyphenol oxidase, which stimulates the mushroom's discoloration. Since the gene editing procedure did not implant foreign DNA, the USDA decided that genetic modification by CRISPR using only native genes does not make the crop a GMO, and they would not regulate the mushroom as one. This set an important precedent that crop companies can modify existing genes in plants and avoid regulation, so long as they do not add genes from other species in the process. This is a boon for plant biologists and might also lead to great commercial reward if Yang or another party were to sell his genetically modified mushrooms since fruits and vegetables genetically engineered to maintain their color have a longer shelf life than their nongenetically altered counterparts—allowing grocery stores to restock less frequently and reap cost-saving benefits [23].

Finally, genetic modification can be used to disrupt disease transmission. Every year, mosquitos infect millions of people with malaria, and every year hundreds of thousands of those individuals die. The World Health Organization stated that in 2015 alone 214 million people were infected with malaria and almost half a million of those patients perished. However, scientists at Harvard University in Cambridge, Massachusetts, plan to use CRISPR technology to genetically alter mosquitos in order to both prevent them from being able to infect people with malaria and limit the population size of malaria-transmitting mosquitos. Scientists have developed a method to control populations called a gene drive. To construct a gene drive, scientists co-opt what are called "selfish genetic elements" to ensure that greater than 99% of a parent's offspring have a copy of the gene. This gene could be a gene that limits fertility, or one that limits the ability of a mosquito to carry a particular disease. Selfish elements work by copying themselves from one parent's chromosome to the other in a

---

Did You Know?

Pretty soon, a three-parent human will not be a myth. No, not a child born of a tryst with the mailman and passed off as another, not a child born of two women in need of a sperm or two men in need of an egg, but an actual human with THREE biological parents.

Here's why: mitochondrial disease is one that affects 1 in 4000 in the United States. While rare, its effects are sometimes devastating. Mitochondria are the "powerhouses" of every cell in the human body except red blood cells and manufacture the energy, or ATP, for the cell. In people with mitochondrial disease, these mitochondria are damaged (mutated), causing everything from muscle weakness and loss of coordination to heart, liver, and kidney disease to complete dementia. There are few treatments that provide help for these children and no cure.

But now there is an opportunity to avoid mitochondrial disease. Thanks to intensive research at Newcastle University and a law passed in Britain in 2016 (home to ~100 children each year with mitochondrial disease), it will become the first country to allow babies to be born using mitochondrial donation, resulting in children made up of three genetic sets instead of two [28].

Here's how it works: all mitochondria are inherited from the mother. Researchers can take out the nucleus from a fertilized egg (from a mother (parent 1) and a father (parent 2)) enabled by *in vitro* fertilization (IVF). The nucleus is then inserted into a non-fertilized donated egg from a healthy mitochondrial DNA woman (person 3) whose nucleus has been removed. Parent 3 contributes only 0.2% of the total DNA, as DNA from the nucleus makes up 99.8% of the genes of a human. But this 0.2% carries those mitochondrial DNA! This now healthy fertilized egg can be implanted into parent 1, disease-free [29].

While opposition to this option for ethical and religious reasons is still in hot debate, we have reached a scientific and a technological option for mitochondrial disease carriers to bear healthy children.

So, the next time you hear someone has three parents, it's not a soap opera. It's science.

---

fertilized egg such that only one parent is represented in that gene in the new baby organism and in all subsequent offspring. This gene would spread through the population and eventually alter the genetic makeup of a huge swath of the species' population. While this endeavor is not yet fully operational, it presents a valuable and lifesaving opportunity to employ CRISPR in the future [26]. Moreover, organizations are doubling down on this tremendous potential. For example, in late 2016 the Gates Foundation dramatically increased its investment in the development of gene drives for mosquitos to a total of US$75 million [27].

Genetic engineering has the potential to present enormous agricultural, medical, and health benefits, but with those benefits comes significant risk. The current regulation around GMO does not impact organisms that have undergone genetic engineering using CRISPR technology because CRISPR does not require the introduction of foreign DNA into an organism. As a result, many scientists are urging for new regulatory structure and limits.

As of 2013, regulators from Canada's Plant and Biotechnology Risk Assessment Unit were arguing that transposition sometimes occurs naturally when new genes are introduced during routine fertilization and breeding and that it would not be unreasonable to expect similar unintended and potentially harmful transposition when scientists intervene to genetically engineer plants or other organisms. Conversely, scientists Williams Price and Lynn Underhill have produced studies that show that genetically modified plants have not had negative, unintended compositional changes, and as such the additional scrutiny typically required for genetically altered plants is unnecessary. Widespread disagreement exists around both the regulatory structure and the limits for these new GMO. While the jury is still out on the exact risks of genetic modification, this debate makes clear how challenging it will be to determine regulatory limits [26]. Also in 2013, the Institute of Medicine, part of the National Academy of Sciences, called for a new group to replace the Recombinant DNA Advisory Committee in order to better regulate the potentially dangerous but also lifesaving clinical research hopefully to come [30].

A number of prominent scientific and bioethics communities are producing reports on the scientific and ethical implications of gene editing, in particular on the advanced CRISPR genetic engineering tool. At the 2015 International Summit on Human Gene Editing held at the National Academy of Sciences, bioethicists, philosophers, scientists, and concerned citizens alike raised concerns about the moral challenges of genetic engineering. While the only consensus raised was one for caution and regulation, different parties argued over the ethical and biological implications of genetically modifying a human embryo versus a human sperm or egg [31]. In addition, in the fall of 2016, the Nuffield Council on Bioethics released a report stating that both human embryos and livestock as subjects for genetic engineering need more in-depth study.

The US National Academies of Sciences, Engineering, and Medicine are also expected to release a major report in 2017 on standards for genetic engineering of human cells. In addition, a team of ethicists based in Europe plan to convene an independent European group to determine guidelines and protocol for CRISPR before it is employed in a medical context [32]. All of this information shows how powerful gene editing is and how far reaching its impact can be. However, there are actually a number of engineering issues that limit our ability to safely use gene editing in humans. As noted earlier in this chapter, CRISPR is known to have issues with accuracy. Scientists have long known of "off-target effects," in which compounds targeting a particular sequence of DNA end up binding an unintended target. Even without a perfect match, this is enough to disturb the original experiment. In the case of CRISPR, this can lead to DNA being cut in unintended areas, potentially disturbing other genes crucial to a healthy life and survival.

Additionally, science has a long way to go before we truly understand the implications of changing one gene. That single gene may cause a ripple effect, like how killing a predator may lead to an overabundance of pests that further upsets an ecosystem. It may be generations before we understand the consequences of changing one gene—and while it is fairly easy to test in quick breeding crops and livestock, it is much more difficult to assess in long-lived human beings.

Finally, as noted earlier, there are countless diseases that are influenced by a variety of genes. It is much simpler to address a single gene-driven disease like cystic fibrosis versus diseases with many possible and highly complicated genetic backgrounds like autism. Modifying multiple genes greatly increases the risk of unintended consequences. In sum, our understanding of gene editing technology and the complexity of biology means that we have a long road ahead of us in curing all genetic disease. Though the implications to create a true designerome[1] for our next populations are huge, genome editing does not necessarily imply that terrible or wonderful futures await. Humans have made unthinkably powerful scientific advances for millennia, and we have built incredible civilizations with these tools. CRISPR can be another gift for us to improve humanity if harnessed and regulated in the right way.

References

1 Kim YG, Cha J, Chandrasegaran S. Hybrid restriction enzymes: zinc finger fusions to Fok I cleavage domain. Proc Natl Acad Sci U S A. 1996 Feb;93(3):1156–60.
2 Adapted from: Yin H, Kanasty RL, Eltoukhy AA, Vegas AJ, Dorkin JR, Anderson DG. Non-viral vectors for gene-based therapy. Nat Rev Genet. 2014;15:541–55.
3 Adapted from: Lewis T. We all kind of marvel at how fast this took off [Internet]. Business Insider; Dec 21, 2015 [cited Dec 1, 2016]. Available from: http://www. businessinsider.com/how-crispr-is-revolutionizing-biology-2015-10
4 Nemudryi AA, Valetdinova KR, Medvedev SP, Zakian SM. TALEN and CRISPR/Cas genome editing systems: tools of discovery. Acta Nat. Jul–Sep 2014;6(3):19–40.

---

1 A "Designerome"(pronounced Designer-Ome) is a currently fictitious genome artificially crafted with the exact specifications of its creator to "design" a super human. I coined the term to reflect the current trajectory of our science and the potential in the future to create perfectly tailored humans, a potential we will all continue to debate the consequences of.

5 Ran FA, Cong L, Yan WX, Scott DA, Gootenberg JS, Kriz AJ, et al. In vivo genome editing using Staphylococcus aureus Cas9. Nature. Apr 9, 2015; 520(7546):186–91.

6 Long C, McAnally JR, Shelton JM, Mireault AA, Bassel-Duby R, Olson EN. Prevention of muscular dystrophy in mice by CRISPR/Cas9—mediated editing of germline DNA. Science. Sep 5, 2014;345(6201):1184–88.

7 Office of Communications and Public Liaison. Muscular dystrophy: hope through research [Internet]. National Institute of Neurological Disorders and Stroke; Aug 2013 [cited Dec 1, 2016]. Available from: https://www.ninds.nih. gov/Disorders/Patient-Caregiver-Education/Hope-Through-Research/Muscular-Dystrophy-Hope-Through-Research

8 Cyranoski D. Chinese scientists to pioneer first human CRISPR trial. Nat News. Jul 28, 2016;535(7613):476.

9 Qasim W, Amrolia PJ, Samarasinghe S, Ghorashian S, Zhan H, Stafford S, et al. First clinical application of TALEN engineered universal CAR19 T cells in B-ALL. Blood. Dec 3, 2015; 126(23):2046.

10 Adapted from: Kochenderfer JN, Rosenberg SA. Treating B-cell cancer with T cells expressing anti-CD19 chimeric antigen receptors. Nat Rev Clin Oncol. May 2013;10:267–76.

11 Raphael J. First ever CRISPR gene-editing trial on human to begin in China [Internet]. Nature World News; Jul 25, 2016 [cited Dec 1, 2016]. Available from: http://www.natureworldnews.com/articles/25743/20160725/first-crispr-gene-editing-trial-human-begin-china.htm

12 ACOG Committee on Genetics. ACOG Committee opinion no. 442: preconception and prenatal carrier screening for genetic diseases in individuals of Eastern European Jewish descent. Obstet Gynecol. Oct 2009;114(4):950–3.

13 Surveillance [Internet]. International Federation of Fertility Societies; [cited Oct 20, 2016]. Available from: http://www.iffs-reproduction.org/?page=Surveillance

14 Controversial genetic tests: German Parliament allows some embryo screening [Internet]. Spiegel Online; Jul 7, 2011 [cited Oct 20, 2016]. Available from: http://www.spiegel.de/international/germany/controversial-genetic-tests-german-parliament-allows-some-embryo-screening-a-773054.html

15 Mosher DS, Quignon P, Bustamante CD, Sutter NB, Mellersh CS, Parker HG, et al. A mutation in the myostatin gene increases muscle mass and enhances racing performance in heterozygote dogs [Internet]. PLoS Genet; May 25, 2007 [cited Oct 20, 2016]. Available from: http://journals.plos.org/plosgenetics/article?id=10.1371/journal.pgen.0030079

16 Regalado A. First gene-edited dogs reported in China [Internet]. MIT Technology Review; Mar 15, 2016 [cited Dec 1, 2016]. Available from: https://www.technologyreview.com/s/542616/first-gene-edited-dogs-reported-in-china/

17 Cyranoski D, Reardon S. Chinese scientists genetically modify human embryos [Internet]. Nature; Apr 22, 2015 [cited Dec 1, 2016]. Available from: http://www.nature.com/news/chinese-scientists-genetically-modify-human-embryos-1.17378

18 Eck A. Scientists use CRISPR to grow human organs inside of pigs [Internet]. PBS; Jun 6, 2016 [cited Dec 1, 2016]. Available from: http://www.pbs.org/wgbh/nova/next/body/scientists-use-crispr-to-grow-human-organs-inside-of-pigs/

19 Reardon S. New life for pig-to-human transplants [Internet]. Nature; Nov 10, 2015 [cited Dec 1, 2016]. Available from: http://www.nature.com/news/new-life-for-pig-to-human-transplants-1.18768

20 Brouillette M. Scientists breed pigs resistant to a devastating infection using CRISPR [Internet]. Sci Am; Feb 4, 2016 [cited Dec 1, 2016]. Available from: https://www.scientificamerican.com/article/scientists-breed-pigs-resistant-to-a-devastating-infection-using-crispr/

21 Ryder D. Infographic: climate change vs. GMOs: comparing the independent global scientific consensus [Internet]. Genetic Literacy Project; Jul 8, 2014 [cited Dec 5, 2016]. Available from: https://www.geneticliteracyproject.org/2014/07/08/climate-change-vs-gmos-comparing-the-independent-global-scientific-consensus/

22 Charles D. In a grain of golden rice, a world of controversy over GMO foods [Internet]. NPR; Mar 7, 2013 [cited Dec 1, 2016]. Available from: http://www.npr.org/sections/thesalt/2013/03/07/173611461/in-a-grain-of-golden-rice-a-world-of-controversy-over-gmo-foods

23 Waltz E. Gene-edited CRISPR mushroom escapes US regulation [Internet]. Nature; Apr 14, 2016 [cited Dec 1, 2016]. Available from: http://www.nature.com/news/gene-edited-crispr-mushroom-escapes-us-regulation-1.19754

24 Adapted from: Bonser K, Layton J. How designer children work [Internet]. How Stuff Works; Ma 10, 2001 [cited Oct 20, 2016]. Available from: http://science.howstuffworks.com/life/genetic/designer-children2.htm

25 PGD—genetic testing of the embryo [Internet]. LifeInvitro; [cited Oct 20, 2016]. Available from: http://www.lifeinvitro.com/pgd.html

26 Powledge TM. Can we regulate gene editing without killing it? [Internet]. Genetic Literacy Project; Aug 1, 2014 [cited Dec 5, 2016]. Available from: https://www.geneticliteracyproject.org/2014/07/29/can-we-regulate-gene-editing-without-killing-it/

27 Regalado A. Bill Gates doubles his bet on wiping out mosquitoes with gene editing [Internet]. MIT Technology Review; [cited Oct 20, 2016]. Available from: https://www.technologyreview.com/s/602304/bill-gates-doubles-his-bet-on-wiping-out-mosquitoes-with-gene-editing/

28 Hogan E. Baby leaps into a brave new world [Internet]. The Economist; Nov 2, 2015 [cited Dec 1, 2016]. Available from: http://www.theworldin.com/article/10461/baby-leaps-brave-new-world

29 About mitochondrial disease—mito FAQ [Internet]. Mito Action; [cited Oct 20, 2016]. Available from: http://www.mitoaction.org/mito-faq

30 Powledge TM. Scientists urge revamped regulations for genetic engineering [Internet]. Genetic Literacy Project; Jan 6, 2015 [cited Dec 1, 2016]. Available from: https://www.geneticliteracyproject.org/2015/01/06/scientists-urge-revamped-regulations-for-genetic-engineering/

31 Travis J. Inside the summit on human gene editing: a reporter's notebook [Internet]. Science; Dec 4, 2015 [cited Dec 1, 2016]. Available from: http://www.sciencemag.org/news/2015/12/inside-summit-human-gene-editing-reporter-s-notebook

32 Ledford H. UK bioethicists eye designer babies and CRISPR cows [Internet]. Nature; Sep 30, 2016 [cited Dec 1, 2016]. Available from: http://www.nature.com/news/uk-bioethicists-eye-designer-babies-and-crispr-cows-1.20713

16

Precision Medicine around the World

China

I arrived at the Prince of Wales Hospital on a 90°F muggy June day right outside of Hong Kong. Not following my coworker's strict orders where to meet him, I wandered directly into the general intake area—bustling with so many people, occasional signs in English but mostly not, and an overall buzz of doctors, nurses, caregivers, children, and parents—that, paired with the heat and my never-ending jet lag, made me queasy enough to want to sit down. But once past the foyer, having found my coworker and stepped into the real hospital, the overall orderliness suggested a structure of a hospital that has positively settled into its population. The hospital opened in 1984 and is the public hospital and teaching hospital of the Faculty of Medicine of the Chinese University of Hong Kong (CUHK). With approximately 1500 beds total, it offers emergency, inpatient, and outpatient services with a large specialized cancer department (the Sir Yue-kong Pao Cancer Centre, opened in 1994). Sir Yue-kong Pao Cancer Centre works closely with the Hong Kong Cancer Institute at the CUHK, which operates three floors of lab space totaling 3500 m² for cancer research. The center's Comprehensive Cancer Trials Unit is the first center in Hong Kong—approved by the National Cancer Institute (NCI) in the United States in 2002—to conduct new NCI drug clinical studies, with investigators in CUHK acting as principal investigators. And CUHK's Department of Pathology also operates out of the Prince of Wales Hospital, receiving 3000–4000 patient samples daily, and accredited by the National Association of Testing Authorities, Australia (NATA), and the Royal College of Pathologists of Australasia (RCPA). Finally, it is home to decorated researcher Dr Dennis Lo, chairman of the Department of Chemical Pathology, director of the Li Ka Shing Institute of Health Sciences, and associate dean (research) and faculty of Medicine, whom I have known for over 20 years from my early days as a scientist. Visiting him in his home base would be the highlight of my day.

While Chinese natives are quick to point out that Hong Kong is not indicative of mainland China, just as Japan is not indicative of the rest of Asia Pacific, the Asia Pacific region, China included, has taken precision medicine by storm

Personalizing Precision Medicine: A Global Voyage from Vision to Reality, First Edition.
Kristin Ciriello Pothier.

China facts

Definition: The People's Republic of China including Hong Kong and Macau

Population (2015): 1.4 billion (2015 UN Est.)

Gross GDP: $11.5 billion (2015 Un Est.)

Cancer incidence: 4,292,000

Map:

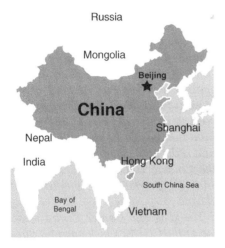

Figure 16.1 China demographic facts.

(Figure 16.1). Compared constantly with the United States with regard to investment and innovation, the regions both have significant and fresh commitment to precision medicine to take care of their specific populations but also in their quest for world leadership. Former US President Obama's announcement at his 2015 State of the Union Address of the launch of a $215 million investment in a new "Precision Medicine Initiative" generated considerable buzz among those both in and outside the life sciences community. This commitment in the President's 2016 budget was to broadly support R&D and other innovation efforts through initiatives by the NIH, FDA, and National Coordinator for Health Information Technology (ONC) [1]. To date, Obama's Precision Medicine Initiative has been the largest federal commitment of US funds to exclusively support precision medicine.

Not to be outdone, however, China revealed its own Precision Medicine Initiative in January 2016, just months after the US plan. Announced by state media as a major national strategic research initiative, China's effort commits up to ¥60 billion (or ~US$9 billion) to precision research initiatives over a 15-year period starting in 2016 [2]. At an average of approximately $600 million per year, the China commitment dwarfs the US plan—which may face changes given new leadership and political shifts.

Against the backdrop of China's current healthcare system, the country's massive commitment to precision medicine makes considerable sense. For example, China's adoption of precision medicine in areas such as oncology is relatively nascent compared with the West despite a high cancer burden. According to the WHO's World Cancer Report, China accounts for 22% (or well over four million) of new cancer cases worldwide, and a disproportionately high 27% of world cancer mortalities [3]. Furthermore, China's 5-year cancer survival rate currently stands at 15%, compared with around 85% in the United States [4]. The lack of access to high-quality diagnostic services and treatments has undoubtedly contributed to these numbers. In China, 0.6 oncologists exists for every 100,000 people. This is similar to Japan's number of oncologists per 100,000 (0.7), but the vast differences between countries with land mass, general health, access to care, and concentrations of people make China's shortage much more dire [5]. As mentioned in another chapter, more than 40% of China's inhabitants live in rural areas, making the shortage of oncologists, most of whom cluster in major cities, even more of an access issue for the country. The Chinese government has acknowledged and pledged to combat this through its recent initiatives such as the one focused on precision medicine. But despite these pledges, the challenges to bringing access and quality up to Western standards appear to be significant.

Despite achieving tremendous economic gains within the last few decades, China's Gini coefficient indicates that between the 1980s and today, China has shifted to a more economically unequal society. More specifically, during that time period, the Gini coefficient increased from 0.3 to 0.5, with some estimates putting it over 0.6. This places China among the most economically inequitable large nations, with just a relatively small number of elite controlling a majority of China's massive wealth [6].

This wealth inequality has trickled down into lack of access to care, as much of China's poorer population is priced out of even relatively basic healthcare services, let alone sophisticated tests and innovative treatments. While reimbursement for the most basic procedures and treatments does exist—covering over 98% of the population as of 2013—current reimbursement schemes leave wide coverage gaps that place a large number of more sophisticated healthcare services and therapies (e.g., genetic testing and targeted cancer drugs) out of reach of the country's poor [7, 8]. Private insurance plans may fill these gaps and are available to those who can afford it, but for most of the country's indigent population, private insurance is simply not an option. In 2014, 34% of

China's over \$500 billion healthcare expenditure was paid out of pocket [9], with a large portion going toward nonreimbursed prescription drugs [7]. Needless to say, China will need to invest heavily on closing the coverage gap on novel treatments in order to bring precision medicine treatments to a majority of its population.

Much like the pace of its economic development, China has also experienced one of the fastest urban transformations in modern history. In 1982, just 20% of the population lived in urban areas, while today that number is around half of its nearly 1.4 billion population. While this shift has brought a significantly larger percentage of the population closer to specialized services, such as advanced healthcare, the nearly 700 million Chinese who are living in rural areas are still geographically isolated from necessary healthcare services relative to their city-dwelling counterparts. From a precision medicine standpoint, we can see the consequences of this unequal distribution of healthcare in figures such as molecular testing rates, which already fall below China's own guidelines nationwide. Detection rate for the lung cancer mutation EGFR, for example, is 51% across China's eight largest urban areas, where patients have greater access to diagnostic services, but only 27% nationwide [8].

The unequal distribution of advanced healthcare services in certain parts of China is compounded by a relative shortage of healthcare professionals within the country. According to WHO estimates, China currently has approximately 1.5 physicians per total 1000 population, which is among the lowest of countries outside of sub-Saharan Africa. By comparison, the density of physicians in the United States is approximately 64% higher, while in some industrialized European countries, this number may be from 2 to 3 times as high [10].

China's shortage of physicians also includes pathologists, who carry out the tests needed to match precision medicine treatments to the appropriate patients. At a recent Chinese pathology industry meeting, Professor Bian Xiuwu, director of the institute from the Third Military Medical University, Southwest Hospital of Pathology, and the branch director of the committee from the Chinese Medical Association of Pathology, commented on this need, saying, "The current pathologist in China is in shortage ... Domestic pathology department needs to improve the equipment configuration in order to meet the requirements of precision medical treatment" [11].

Innovations and alternate approaches such as digital pathology, which could allow for more accurate analysis and interpretation of samples and remote consultations, may help alleviate this shortage of pathologists to some extent. Dr Eric Glassy, president of the Digital Pathology Association and medical director of the Affiliated Pathologists Medical Group in California, notes that "For countries such as China, in which there is a shortage of pathologists as well as laws against sending tissue outside of the country ... digital pathology provides a way to gain access to expert opinion outside of the country" [12].

Another innovation coming right from Dr Lo's labs is noninvasive prenatal testing (NIPT). When testing for chromosomal abnormalities in a fetus, typically women were screened with a protein-based risk test. This test, called the "quad screen," only determines the "risk" that a mother is carrying a chromosomal abnormal fetus; it doesn't definitively diagnose the fetus. Then, if the risk was assessed as high, the patient would move to the invasive amniocentesis test where a needle punctured the amniotic sack to sample the amniotic fluid to definitively determine the fetus's chromosomal state. What Dr Lo and others invented was a way to replace the amniocentesis with a simple blood test on the mother's blood to extract circulating fetal DNA. This fetal DNA is then analyzed to view chromosomal abnormalities such as Down syndrome (trisomy 21). Since its launch a few years ago, it has become the new standard for women in many parts of the world. Indeed, NIPT is offered in China, the United States, most parts of the EU, some parts of LatAm where patients are interested (there is less prenatal testing in general depending on the region's stance on pregnancy termination), and parts of India and the Middle East. Why this is of such interest overall is because while the technology can be used for prenatal care, it can also be used to detect circulating tumor DNA, making the way for potential noninvasive cancer testing without having to sample the tumor directly.

In general, China has doubled down on its commitment to precision medicine in the clinic through its cancer centers, and its research through academic institutions like CUHK and others, and through its Precision Medicine Initiative. The sheer amount of funding and diversity of organizations involved have positioned China to become a future leader in the field. Specifically, the publicly supported Precision Medicine Initiative will engage both public research organizations and private businesses. For example, institutions such as the Sichuan University West China Hospital, Tsinghua University, and Fudan University (whose affiliated hospitals include Huashan Hospital in Shanghai, which provides clinical services for over 20,000 inpatients annually in addition to 1.5 million outpatient and emergency visits each year and is known for its progressive and prestigious staff and offerings) plan to collaborate on establishing a precision medicine center that will sequence one million human genomes. Private collaborators are expected to provide a technical platform and add-on services for the effort [2].

In a separate project, telecom company Huawei and leading Chinese CRO, WuXi AppTec, plan a partnership to develop a cloud-based platform for the Precision Medicine Initiative along with sequencing giants BGI and Berry Genomics, which will provide the genomic sequencing services [13]. Involvement of these private players, which are among the largest and most prestigious in China, underscores the commitment and magnitude of China's Precision Medicine Initiative. BGI, for example, is currently the largest genomic sequencing provider in the world, while Berry Genomics—China's second largest sequencing provider—is the sole company in China approved by the China Food and Drug Administration (CFDA) to offer NIPT to any hospital or clinic in the country [14, 15]. NIPT currently makes up a majority of clinical next-generation sequencing (NGS) tests in China, although oncology and rare diseases are also being aggressively pursued [16].

On the reimbursement front, there has been some recent movement on coverage of targeted cancer therapies. For example, a few wealthier localities such as Guangzhou, Shenzhen, and Qingdao have begun experimenting with reimbursement for EGFR-targeted therapies in non-small cell lung cancer (NSCLC), among other targeted therapies. For reimbursement, patients are required to have been diagnosed with NSCLC and tested positive for EGFR mutations at pre-approved hospitals. Screening is only covered by insurance retroactively and only if a patient tests positive for the mutation. If a patient tests negative for an EGFR mutation, the test must be paid for out of pocket [8, 17]. By the end of 2017, however, a new special insurance scheme for severe diseases such as cancer is expected to expand coverage across China, although it is unclear whether this reform will be able to cover the costs of most new precision medicine therapies [7].

Another recent effort to complement the increased focus in precision medicine is China's relatively new research infrastructure around NGS. NGS hit a temporary snag in China in 2014 when Chinese regulatory authorities, due to concerns around safety and quality of domestically manufactured products, banned all NGS-based testing within the country until NGS platforms and tests could be cleared by the CFDA. Just a few months later, in July 2014, the first NGS platform and test kit, manufactured by BGI, was approved in China [18]. Then, in a rapid succession of events, Chinese authorities announced in December 2014 the first seven pilot sites for clinical NGS testing in genetic diseases and NIPT. A month later, Chinese authorities announced 109 sites approved to pilot NGS NIPT clinical testing. And finally in March 2015, the first 20 approved sites to pilot NGS testing in oncology were made official [16, 19]. Since then, NGS has been increasingly embraced and should be a strategic focus of China's Primary Medicine Initiative in the coming years.

In just a handful of years, China has made advancements within not only precision medicine but also its healthcare system as a whole, and companies are popping up to capitalize on the portion of the Chinese population who has the interest and the money to spend more time on their health. Companies, for example, like Prenetics—a data-based wellness company who offers lab tests for drug metabolism and nutrition paired with a data-rich output guiding the patient to a healthier lifestyle—are becoming more popular and more relevant to a larger percentage of the market today. But as I climbed the old, never-ending stairs upon stairs on Ladder Street in Hong Kong trying to get to the historic Tung Wah Hospital (established in the 1870s at the beginning of the third bubonic plague) and instead finding the Man Mo Temple, or reviewed the continuum of thousands of hospitals and surrounding provincial environments to visit for Shanghai coming up, I was struck by how far we have yet to go to transition China's precision medicine offerings in a more standardized way throughout its cities. Systemic challenges such as inequality, quality standards, and the distribution of healthcare will likely persist as significant barriers for many years; however, China, with its abundance of resources, recent commitments, and eye toward the future, should remain a leader in the future of precision medicine.

References

1 FACT SHEET: President Obama's precision medicine initiative [Internet]. Whitehouse.gov; Jan 30, 2015 [cited Jan 4, 2017]. Available from: https://www.whitehouse.gov/the-press-office/2015/01/30/fact-sheet-president-obama-s-precision-medicine-initiative

2 China embraces precision medicine on a massive scale: Nature News & Comment [Internet]; Jan 6, 2016 [cited Jan 3, 2017]. Available from: http://www.nature.com/news/china-embraces-precision-medicine-on-a-massive-scale-1.19108

3 Chen W, Zheng R, Baade PD, Zhang S, Zeng H, Bray F, et al. Cancer statistics in China, 2015. CA Cancer J Clin. Mar 1, 2016;66(2):115–32.

4 Stewart BW, Wild CP (eds.). World Cancer Report 2014 [Internet]. IARC Publications; [cited Jan 3, 2017]. Available from: http://publications.iarc.fr/Non-Series-Publications/World-Cancer-Reports/World-Cancer-Report-2014

5 Garfield DH, Brenner H, Lu L. Practicing western oncology in Shanghai, China: one group's experience. J Oncol Pract [Internet]. 2013;9(4); [cited Apr 11, 2017]. Available from: https://www.ncbi.nlm.nih.gov/pmc/articles/PMC3710181/

6 Inequality in China: up on the farm [Internet]. The Economist; May 14, 2016 [cited Jan 4, 2017]. Available from: http://www.economist.com/news/finance-and-economics/21698674-rising-rural-incomes-are-making-china-more-equal-up-farm

7 Mossialos E, Wenzl M, Osborn R, Sarnak D (eds.). 2015 international profiles of health care systems [Internet]. The Commonwealth Fund; Jan 2016 [cited Jan 3, 2017]. Available from: http://www.commonwealthfund.org/~/media/files/publications/fund-report/2016/jan/1857_mossialos_intl_profiles_2015_v7.pdf

8 靶向药在中国:规范治疗是难题_医生专访_39健康网 [Internet]. Jbk.39.net; 2016 Mar 1 [cited 2017 Jan 3]. Available from: http://jbk.39.net/fa/yszf/160301/4777431.html

9 Health expenditure per capita (current US$) [Internet]. The World Bank; [cited Jan 4, 2017]. Available from: http://data.worldbank.org/indicator/SH.XPD PCAP

10 Density of physicians (total number per 1000 population): latest available year [Internet]. World Health Organization; [cited Jan 3, 2017]. Available from: http://gamapserver.who.int/gho/interactive_charts/health_workforce/ PhysiciansDensity_Total/atlas.html

11 The first China digital pathology summit ringing the construction of digital pathology horn for the first time [Internet]. UNIC-Healthcare; Mar 11, 2016 [cited Jan 3, 2017]. Available from: http://www.unic-tech.com/english/news/ companyNews/pic/20160311/firstDigitalPathCon.jsp

12 Nierengarten MB. Pathology in the digital era [Internet]. ENTtoday; Aug 9, 2016 [cited Jan 3, 2017]. Available from: http://www.enttoday.org/article/ pathology-digital-era/2/

13 Huawei, WuXi AppTec to provide cloud platform for China precision medicine initiative [Internet]. GenomeWeb; Mar 17, 2016 [cited Jan 3, 2017]. Available from: https://www.genomeweb.com/informatics/huawei-wuxi- apptec-provide-cloud-platform-china-precision-medicine-initiative

14 Trans-omics for a better life [Internet]. BGI; [cited Jan 4, 2017]. Available from: http://www.genomics.cn/en/index

15 贝瑞和康 | 北京贝瑞和康生物技术股份有限公司 [Internet]. Berry Genomics; [cited Jan 4, 2017]. Available from: http://www.berrygenomics.com/

16 Heger M. Clinical NGS market in China poised to take off as China FDA looks to establish guidelines [Internet]. GenomeWeb; Sep 18, 2015 [cited Jan 3, 2017]. Available from: https://www.genomeweb.com/sequencing-technology/ clinical-ngs-market-china-poised-take-china-fda-looks-establish-guidelines

17 11种靶向药纳入深圳重疾补充医保报销目录---深圳晚报 [Internet]; [cited Jan 3, 2017]. Available from: http://wb.sznews.com/html/2015-11/04/ content_3377167.htm

18 CFDA approved next generation sequencing diagnostic products [Internet]. New Center; Jul 2, 2014 [cited Jan 4, 2017]. Available from: http://www. genomics.cn/en/news/show_news?nid=100050

19 卫计委第一批高通量测序技术临床应用试点单位(附名单) - 遗传 - 艾兰博曼 医学网—从检验到临床,从临床到检验 [Internet]; [cited Jan 3, 2017]. Available from: http://www.alabmed.com/content-147-13970-1.html

17

A New Hope

The Future of Precision Medicine

> Precision medicine diagnostics represent one of the largest investment arbitrage opportunities in healthcare. They allow us to access the genetic code for maintaining a healthy human being against metabolic, lifestyle and environmental changes. They are the gateway to enormous downstream cost control and successful application of treatments. We are very bullish on diagnostics and precision medicine overall for the short and long term.
> —*James McCullough, Partner, Renwick Capital*

If you're still with me, I hope you'll agree that precision medicine is here to stay—particularly in the world of cancer. In fact, for the foreseeable future, oncology is likely to remain the vanguard of innovation in personalized health-care, and the past two decades of cancer treatment have demonstrated the value of the right-drug/right-patient/right-time approach. The prominence of cancer within the precision medicine world will be driven, also, by market forces. Oncology is a vast market for pharmaceuticals, with over $80 billion worth of cancer drugs sold worldwide in 2015 alone [1]. As a result, oncology treatments enabled by precision medicine are likely to keep featuring promi-nently on companies' balance sheets—and the incentives to keep developing them will continue to be strong.

Precision oncology owes its future to much more than just history and economics, though. The current decade has seen a series of breakthroughs in a new category of treatments hitting the market today, collectively referred to as immuno-oncology, which are redefining what it means to be a successful therapy in several particularly deadly cancers—and giving patients much-needed cause for hope. To date, these therapies mostly impact only small subsets of patients, but a tidal wave of clinical trials is underway to find more and more cases in which they might be of use. Importantly, too, the application of precision medicine to immuno-oncology treatments is in its infancy—mean-ing, as we have seen, that the ability to target the right patients for the right drug will only improve from here.

Personalizing Precision Medicine: A Global Voyage from Vision to Reality, First Edition.
Kristin Ciriello Pothier.
© 2017 John Wiley & Sons, Inc. Published 2017 by John Wiley & Sons, Inc.

Immuno-oncology

As a general definition, an immuno-oncology treatment is any therapeutic approach—a drug or, as we'll see, a vaccine or cell-based therapy—that musters the body's own immune system to attack a tumor. The treatment doesn't harm the tumor directly; instead it organizes, empowers, or directs normally existing immune cells and processes to do the damage. They work like a laser-guided precision bomb, a warhead that finds its target by following the beam of (otherwise harmless) laser light being shone on it. T lymphocytes, or T cells, the immune system's most sophisticated and adaptable system of defense, are the bomb; immune therapies are the laser.

The basic idea isn't a new one, and the recent crop of immune therapies—beginning with Yervoy, launched by Bristol-Myers Squibb (BMS) in 2011—isn't actually the first to be available. In fact, immune therapies for cancer first appeared in the 1980s and 1990s when two substances, called interferon and interleukin-2 (or IL-2), were approved to use in cancer. Neither, strictly speaking, is a drug; both are proteins that are ordinarily present in the body anyway. They belong to a class of proteins called *cytokines,* which are collectively a group of signaling molecules that circulate in the bloodstream. A number of cytokines are involved in controlling the immune system—activating or deactivating various components of it—and the role of both interferon and IL-2 is to turn on, in a general way, the inflammatory and other defensive responses that respond to foreign organisms (like bacteria) in the body. These are the same groups of cells that respond to tumors, so you see the logic in using them for treating cancer. The problem, though, is that they're *too* general: while they raise the overall level of activity in the immune system, they don't offer any assistance in targeting the tumor specifically. As a result, they were highly effective in only a small fraction of patients. Worse, they were toxic: a state of continual, generalized immune response is hard on a number of end organs in the body. (Both interferon and IL-2 are related to, and often associated with, interleukin-1, which is called the "endogenous pyrogen"—it's what gives you a fever. This is why patients taking interferon report feeling like they have the flu, possibly for months at a time.)

In contrast to the 1980s technology of interferon and IL-2, the current crop of immune therapies strikes an appealing balance. They don't attack the body like cytokine treatment; nor, however, do they target mutations—like EGFR in lung cancer—that may be present in only a small proportion of patients. They're tumor-specific, but generalizable—mostly—among patients. This is the essence of the breakthrough and, to again reference one of my favorite *Star Wars* movies, *A New Hope* in our personalized arsenal against cancer.

Current-generation immuno-oncology treatments take a number of forms and work by way of a number of mechanisms. What they have in common is that they were potentiated by recent advances in basic science. One major field of study has been the tumor microenvironment—the area of the body immediately adjacent to a tumor, which has proven to be teeming with a melting pot of cells in constant microsecond communication with the tumor and with each other. Another has been the fundamental biology of many of these cell types regardless of where they are in the body—including before they're activated and sent to the tumor. Thanks to these studies, there are now at least six approaches to treating a tumor with immuno-oncology (Figure 17.1) [2]. These approaches fall roughly in chronological order, with respect to the processes you'd want to see happen in an effective immune response.

First, you have to get the cells you need into the tumor microenvironment using chemokines—a type of signaling molecule involved in localizing cells to particular regions of the body. You also need to be sure they're differentiated. Most cells arrive at their final active state by starting from a stem cell and becoming progressively more specific types of cell—a process called "differentiation" that is driven by other molecules in the body. In other words, you want T cells (for instance) both to be in the tumor microenvironment *and* to have differentiated into one of the subtypes of T cells that will respond effectively to the tumor.

Second, the immune cells, or *leukocytes*, need to be primed to respond. Vaccines are a classic example of priming, not just in oncology but in infectious diseases: immune cells are conditioned to recognize a given type of bacteria or virus or tumor cell, such that they recognize it as hostile and immediately begin to respond in force. Vaccines activate what's known as the *adaptive* immune response, which is the portion of your immunity that's specific to a particular target. With adaptive immunity, a full response, including antibodies, is available almost immediately for any invader you've been exposed to before versus the four or more days it might otherwise take if the body hadn't already been exposed to the pathogen. This is why you can't get chicken pox twice, or even the same strain of the cold virus. (Don't get too excited—there are at least a hundred cold viruses so you will keep getting them.) To complicate things slightly, the innate immune response—the portion of the immune system that isn't specific to a particular pathogen—can also be primed to respond to tumors.

This same innate immune response—including macrophages (or monocytes), the body's cleaning crews, which digest debris and even whole organisms—can be not only primed but also activated. They're not just looking for an invader; once activated, their claws are out.

The fourth phase of the response, *secondary* co-stimulation, is a little more nuanced. T cells, once activated, remain activated for a matter of days—and then run out of steam. Secondary co-stimulation gives them an added boost to keep attacking tumors while also multiplying to add to their force. (The "co-" here, by the way, refers to the fact that other signaling molecules are likely working toward the same end.)

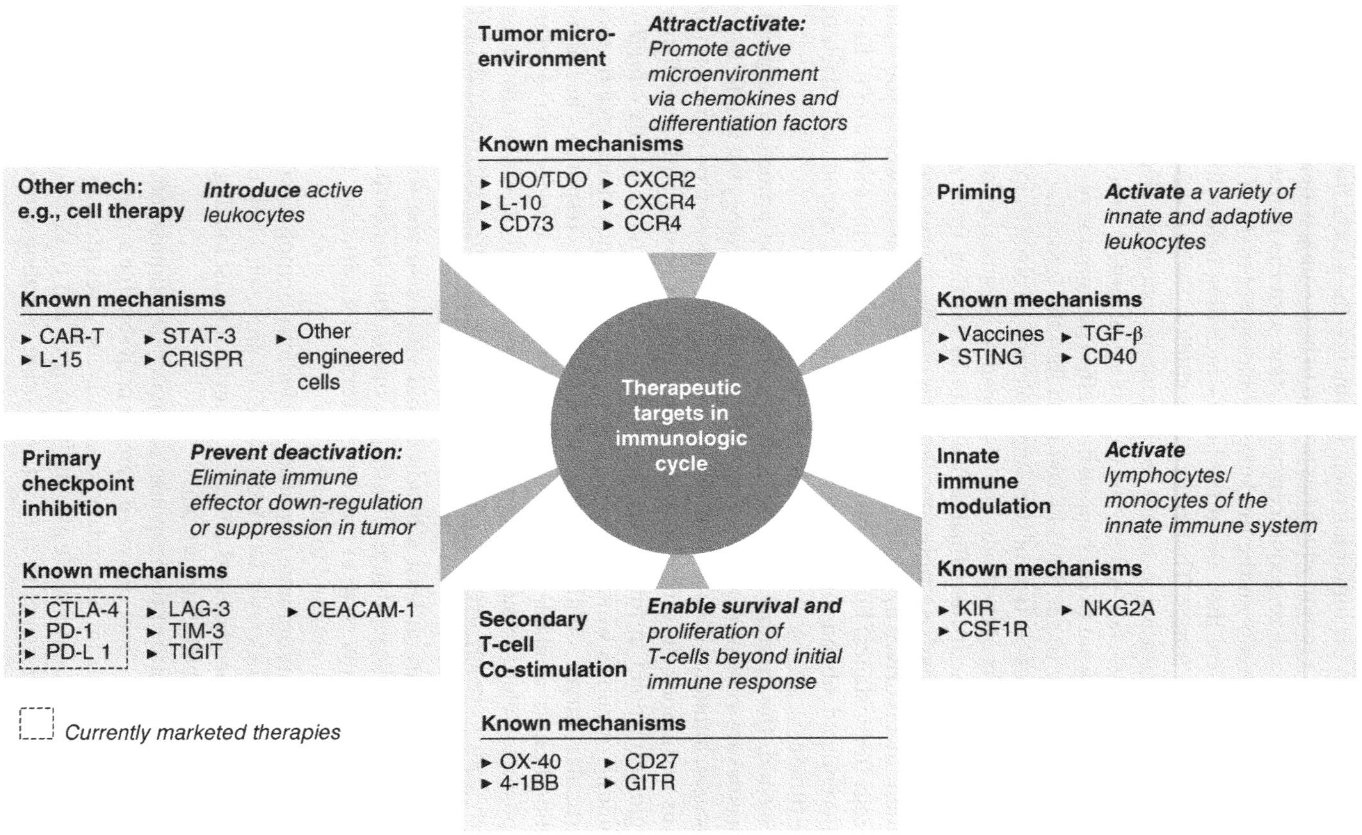

Figure 17.1 Therapeutic targets in immuno-oncology [2].

The fifth phase of a counter-tumor immune response is critical—because of a particularly devious adaptation that many tumors are capable of. The immune system, as with so many other systems and processes in the body, needs to be limited, held in balance with the rest of our system. As we've seen, interferon and IL-2, which wind up the immune system almost indiscriminately, do a great deal of harm as well as good. And so even T cells, which lead the majority of the attack on tumors, need to be deactivated when they're no longer needed. They're deactivated by means of a receptor on their surface called PD-1, where PD stands for "programmed death." When its ligand—the protein that binds PD-1 and is called, suitably, PD-L1—locks into place, the cell initiates its self-destruct sequence, apoptosis. (In addition to PD-1, a second receptor, called CTLA-4, needs to be bound as well.) Typically, PD-L1 appears on the surface of other cells that are involved in restoring balance and shutting down the immune response once the threat is over. This is why the process is known as an *immune checkpoint*. However, this is an Achilles heel that tumors, unfortunately, are able to exploit. In fact, some of the most dangerous, fastest-spreading tumors, like metastatic melanoma, express PD-L1, which leaves them virtually untouchable.

As with the other immune pathways we've reviewed so far, PD-1/PD-L1 signaling is a focus of tremendous research. There's one important difference though: PD-1/PD-L1 treatments, known as *checkpoint inhibitors*, already exist. Four drugs—Yervoy (which targets CTLA-4), Opdivo (which targets PD-1), Keytruda (PD-1), and Tecentriq (PD-L1)—are available for prescription and are used in a widening range of deadly cancers, including cancers of the skin, kidneys, and lungs. When they're effective, they can offer a longer life—sometimes a much longer life—to patients who may otherwise have had little time left.[1]

So far, these treatments are still a relatively small corner of the cancer market—roughly $3 billion out of the overall $80 billion worldwide [1]. However, their success so far has encouraged their inventors to explore any and every cancer where there even *may* be a rationale for treatment. They even provide a kind of halo effect, strengthening the rationale of other treatments that aim to marshal the immune system to fight tumors. The result that pharmaceutical analysts expect is immense. By 2020, immuno-oncology treatments—led by checkpoint inhibitors, but not exclusively—are likely to grow from about 4% of the market to 15%. As the overall cancer market grows (from $80 to $142 billion), immune therapies may reach nearly $26 billion in sales. That's 54% growth *every year*. Much more importantly, though, these treatments, based on what we know today, have the potential to touch millions of lives.

---

1 The remaining category of treatments—which includes therapies based on cell and gene therapies—won't be a focus here; the field is changing so rapidly that anything I write could be old news tomorrow.

Making that prediction a reality requires an immense effort in R&D. In the fall of 2016, 6 companies (BMS, Merck (MRK), Roche (RHHBY), AstraZeneca (AZN), Novartis (NVS), and Pfizer (PFE)) have 106 clinical programs—a combination of a treatment and a single disease—in immuno-oncology (Figure 17.2). (Note that this covers only "late-stage" programs—phase II and phase III clinical trials where the treatment has already been shown to be safe and is being studied for efficacy.)

Checkpoint inhibitors are a major focus of attention—but far from the only one. The oncology pipeline overall is robust and diverse, as shown in Figure 17.3 [3].

As you can see, drugs in development fall into three major categories: those that target the immune system (like the checkpoint inhibitors), those that are specific to tumor cells (like Herceptin or other targeted therapies), and those that aren't specific but affect tumor cells disproportionately (like interferon and IL-2). What is not demonstrated in this figure, however, is the impact of all of the combinations of these drugs in development. In fact, the next generation of immuno-oncology treatments will likely rely on combination therapy, targeting both the immune checkpoint that deactivates T cells and other pathways, in the hope of a synergistic effect. The prospect of combo therapies, in fact, has provoked an unprecedented level of cooperation within the pharmaceutical industry, even among companies who ordinarily are fierce competitors for each other. Companies are striking deals to develop certain drugs together, in the hope of

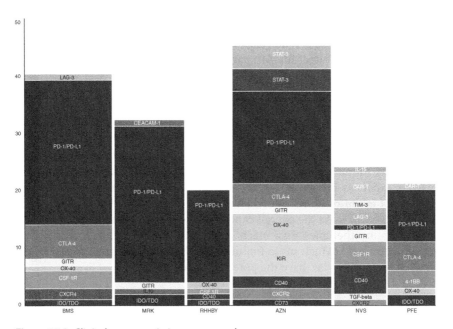

Figure 17.2 Clinical programs in immuno-oncology.

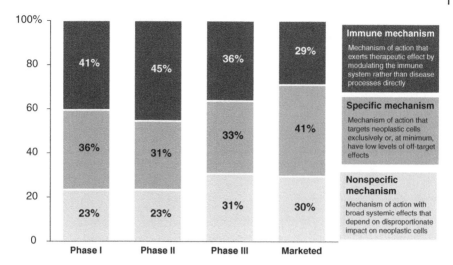

Figure 17.3 Therapeutic pipeline in oncology [3].

creating a powerful combo therapy with a drug they may already have in-house. In effect, it gives companies a larger pipeline without needing to spend any more than they already are on research, and the shared costs reduce everyone's risk [3].

This is why, of the roughly 400 deals struck between companies this decade in oncology, about 250 have been for immune therapies. Remember: these drugs are only about 4% of sales in oncology overall. The investment in the future is immense (Figure 17.4).

In short, immuno-oncology represents a tremendous opportunity for medicine. Looking toward the future, it also represents a tremendous opportunity for *precision* medicine—one that has barely been tested to date. Notably, as intricate as the biology of PD-1 and PD-L1 may have seemed, recent findings suggest that reality is, as usual, far more complex than we thought. Although it might seem obvious that the "right patient" for your checkpoint inhibitor is one whose tumor expresses PD-L1, it doesn't quite work out that way. Patients with more PD-L1 respond better to most of the relevant drugs—but patients with *no* PD-L1 still benefit, which, on the surface, makes no sense. This is why, although companion diagnostics are available for certain checkpoint inhibitors, the value of detecting PD-L1 is a topic of intense debate. The one thing that seems clear is that as our knowledge of the immune system evolves and our ability to manipulate it to save lives evolves too, our diagnostic technology will have to keep up. The potential to do good is vast; in precision medicine, this is what we live for.

This potential will not be realized without all the stakeholders responsible for developing, disseminating, and receiving precision medicine working together to achieve that one common goal. The rise of precision medicine has

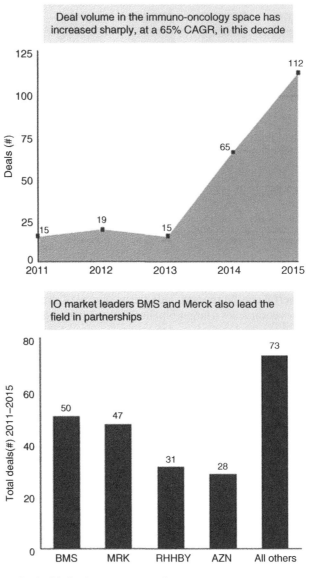

Figure 17.4 Market highlights in immuno-oncology.

called for greater collaboration between an increasingly complex and varied set of stakeholders who need to partner across the life sciences and healthcare landscape and do it worldwide. Every year, over the last 5 years, I have drawn a stakeholder map of players in precision medicine. And every year, the universe of stakeholders gets bigger (Figure 17.5).

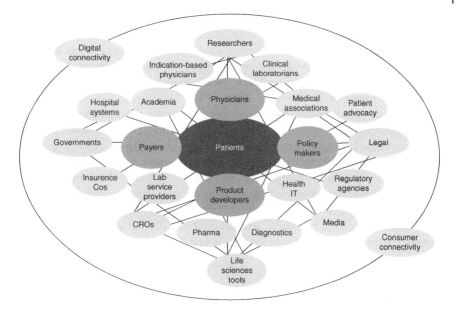

Figure 17.5 Precision medicine stakeholder map in 2017.

In most developed countries, gone are the days of a one-on-one exchange between one doctor and one patient to treat cancer, especially with precision medicine. Indeed, the number of stakeholders necessary to invent, translate, and deliver precision medicine to patients worldwide is ever-increasing, bringing with it a series of structural, cultural, and business-related challenges to overcome. The stakeholder combinations are endless and interconnected. Physicians are interconnecting as oncologists, core laboratorians, pathologists, and psychiatrists to treat cancer in a holistic way. Drug developers and diagnostics developers are finally working more harmoniously together with the common goal of using diagnostics to deliver the RIGHT targeted therapy to a patient. Payers are working with medical institutions, drug and diagnostics developers, and patients themselves to pay for these advances in an economical way. Policy makers are working with regulators to try to streamline processes to ensure both safety and speed to market for these products and services. Patient advocacy groups are crafting and lobbying for more financial and ancillary support for patients, families, and caregivers throughout the patient's cancer journey. And consumer-based and digital-based companies are working with each of these stakeholders to enable information and technology flow to usher in precision medicine of tomorrow. And all of these stakeholders are emerging globally. As highlighted in chapters throughout this book, markets all over the world are clamoring for precision medicine to better treat the infectious, cardiovascular, and oncological diseases of populations at home.

However, the implementation and access challenges differ greatly in each region of the world, stakeholders take on different roles in each region, and there is much work to be done to ensure equal access no matter where you call home. Just as the patient–doctor relationship has changed dramatically, so must the rest of medicine to embrace the diversity of enablers, partnerships, and types of care worldwide.

Throughout this book, I have attempted to detail the past, present, and future of precision medicine for anyone being introduced to the space for the first time. I have balanced the excitement and progress with precision medicine with the major challenges we face as we enter the next generation of patient-empowered, digitally enabled care worldwide. I hope that my goal to "personalize" precision medicine for all of my readers was met, and will catalyze a new wave of interest and involvement from my readers, wherever they are in the world. In precision medicine, we will continue on that global voyage from vision to reality, and indeed, our journey is just beginning.

References

1 EvaluatePharma. World Preview 2016, Outlook to 2022; Sept 2016 [cited May 8, 2017]. Available from: http://info.evaluategroup.com/rs/607-YGS-364/images/wp16.pdf.

2 Adapted from: Chen DS, Mellman I. Oncology meets immunology: the cancer-immunity cycle. Immunity. Jul 23, 2013;39(1):1–10.

3 Adapted from: Palmer S, Kuhlmann G, Pothier K. IO nation: the rise of immuno-oncology. Curr Pharmacogenomics Pers Med. 2014;12(3):176–81.

Afterword

I have changed all of the previously unpublished patient names in this book in order to protect their identities save one, my grandfather, Angelo Ciriello. Grandpa C's battle with metastatic lung cancer defined years of my childhood. If he had been diagnosed today, he would have had many more choices personalized just for him and the cancer he battled.

This picture was taken of him the summer before he died, in his favorite place in the world, on his sailboat with our family. He was always able to skillfully navigate and conquer the waves in front of him, but didn't have the ability back then to navigate and conquer his cancer. He was the patriarch of our family and even now, over 30 years later, we miss him. For all of you who have one or more loved ones with cancer, I encourage you all to take the time to read and share these stories. Take extra time to help someone struggling with understanding a diagnosis to understand it. Through all of my travels worldwide, and all of my conversations with stakeholders, whether they be physicians or payers, business leaders or laboratorians, caregivers or patients, from Boston to Dubai to São Paulo, all have echoed the need for more basic understanding of precision medicine and how to access its power. You may save a life in the process or, as a start, allow those around you to better understand their options for a healthier future.

Figure b.1 Grandpa C.

Kristin Ciriello Pothier
February 2017

Personalizing Precision Medicine: A Global Voyage from Vision to Reality, First Edition.
Kristin Ciriello Pothier.
© 2017 John Wiley & Sons, Inc. Published 2017 by John Wiley & Sons, Inc.

Index

Personalizing Precision Medicine: A Global Voyage from Vision to Reality, First Edition.
Kristin Ciriello Pothier.
© 2017 John Wiley & Sons, Inc. Published 2017 by John Wiley & Sons, Inc.

Printed and bound by CPI Group (UK) Ltd, Croydon, CR0 4YY

27/10/2024

14580273-0002